Collins

Student Support Materials for AQA

AS/A-level Year 1

Chemistry

Paper 2 Organic chemistry and relevant physical chemistry topics

Authors: Colin Chambers, Geoffrey Hallas, Andrew Maczek, David Nicholls, Rob Symonds, Stephen Whittleton

William Collins' dream of knowledge for all began with the publication of his first book in 1819.

A self-educated mill worker, he not only enriched millions of lives, but also founded a flourishing publishing house. Today, staying true to this spirit, Collins books are packed with inspiration, innovation and practical expertise. They place you at the centre of a world of possibility and give you exactly what you need to explore it.

Collins. Freedom to teach

HarperCollins Publishers
The News Building
1 London Bridge Street
London SE1 9GF

HarperCollins Publishers
Macken House
39/40 Mayor Street Upper
Dublin 1, D01 C9W8
Ireland

Browse the complete Collins catalogue at
www.collins.co.uk

10 9 8 7

© HarperCollins*Publishers* 2016

ISBN 978-0-00-818949-5

Collins® is a registered trademark of HarperCollins*Publishers* Limited

www.collins.co.uk

A catalogue record for this book is available from the British Library

Thanks to John Bentham and Graham Curtis for their contributions to the previous editions.

Commissioned by Gillian Lindsey
Edited by Alexander Rutherford
Project managed by Maheswari PonSaravanan at Jouve
Development by Tim Jackson
Copyedited and proof read by Janette Schubert
Typeset by Jouve India Private Limited
Original design by Newgen Imaging
Cover design by Angela English
Printed in Great Britain by Ashford Colour Press Ltd.
Cover image © Shutterstock/isaravut

MIX
Paper | Supporting responsible forestry
FSC
www.fsc.org
FSC™ C007454

Contents

Introduction

To the student

This book covers the Physical chemistry in section 3.1.5 of the AQA specification and the Organic chemistry in sections 3.3.1 to 3.3.6.

Questions in AS Paper 2 will assess your knowledge and understanding of these sections of the specification. This paper will also examine the Physical chemistry in sections 3.1.2 to 3.1.4 and 3.1.6 which were covered in detail in *Collins Student Support Materials for AQA – AS/A-level Year 1 Chemistry: Paper 1 Inorganic chemistry and relevant physical chemistry topics* (978-0-00-818078-2). A table of these Physical chemistry sections is given below.

Specification section	Topic	Page number
3.1.2	**Amount of substance**	**13**
3.1.2.1	Relative atomic mass and relative molecular mass	13
3.1.2.2	The mole and the Avogadro constant	13
3.1.2.3	The ideal gas equation	15
3.1.2.4	Empirical and molecular formula	15
3.1.2.5	Balanced equations and associated calculations	16
3.1.3	**Bonding**	**21**
3.1.3.1	Ionic bonding	21
3.1.3.2	Nature of covalent and dative covalent bonds	22
3.1.3.3	Metallic bonding	23
3.1.3.4	Bonding and physical properties	24
3.1.3.5	Shapes of simple molecules and ions	27
3.1.3.6	Bond polarity	29
3.1.3.7	Forces between molecules	30
3.1.4	**Energetics**	**33**
3.1.4.1	Enthalpy change	33
3.1.4.2	Calorimetry	34
3.1.4.3	Applications of Hess's law	36
3.1.4.4	Bond enthalpies	38

3.1 Physical chemistry

3.1.5 Kinetics

Essential Notes

In addition, before a reaction can occur, it is often necessary for the orientation of molecules to be correct on collision.

3.1.5.1 Collision theory

When two substances react, particles (molecules, atoms or **ions**) of one substance must collide with particles of the other. However, not all collisions result in a reaction; i.e. not all collisions are productive. This situation arises because particles will react only when they collide with sufficient energy. The minimum energy necessary for reaction is known as the **activation energy**. In practice, most collisions do not lead to a reaction because the particles do not collide with sufficient energy. If the activation energy is not reached, particles simply bounce apart unchanged.

3.1.5.2 Maxwell–Boltzmann distribution

In a sample of gas or liquid, the molecules are in constant motion and collide both with each other and with the walls of their container. Such collisions are said to be **elastic**; i.e. no energy is lost during the collision, but energy can be transferred from one molecule to another.

Consequently, at a given temperature, molecules in a particular sample will have a spread of energies about the most probable energy. James Clark Maxwell and Ludwig Boltzmann derived a theory from which it is possible to draw curves showing how these energies are distributed. A plot of the number of molecules with a particular energy against that energy (see Fig 1) is known as the Maxwell–Boltzmann distribution curve.

Fig 1
Distribution of energies at a particular temperature

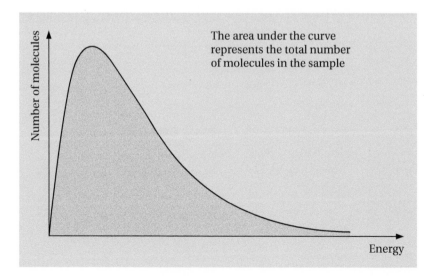

The area under the curve represents the total number of molecules in the sample

This distribution curve has several important features. There are no molecules with zero energy and only a few with very high energies. There is also no maximum energy for molecules – the curve in Fig 1 approaches zero **asymptotically** at high energy. The most probable energy of a molecule corresponds to the maximum of the curve as indicated in Fig 2.

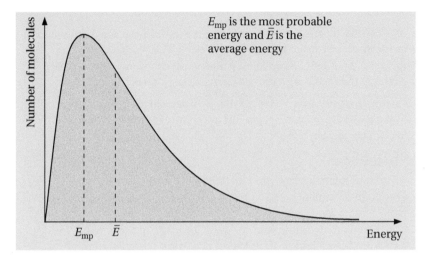

E_{mp} is the most probable energy and $\bar{E}$ is the average energy

Fig 2
Most probable and average energies

Effect of temperature variation on the Maxwell–Boltzmann curve

If the temperature of the sample is increased from T_1 to T_2, the average energy of the molecules increases, and the most probable energy of the molecules increases. The spread of energies also increases and the shape of the distribution curve changes as shown in Fig 3. For a fixed sample of gas, the total number of molecules is unchanged so the area under the curve remains constant (see also *Collins Student Support Materials: A-Level year 1 – Inorganic and Relevant Physical Chemistry*, section 3.1.6).

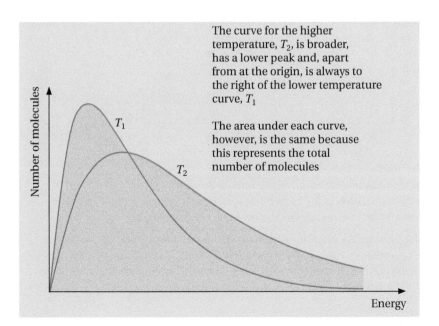

The curve for the higher temperature, T_2, is broader, has a lower peak and, apart from at the origin, is always to the right of the lower temperature curve, T_1

The area under each curve, however, is the same because this represents the total number of molecules

Fig 3
Distribution of energies at two temperatures

3.1.5.3 Effect of temperature on reaction rate

> **Definition**
> The *rate of a reaction* is defined as the change in **concentration** of a substance in unit time.

An increase in temperature always increases the rate of a reaction. According to kinetic theory, the mean kinetic energy of particles is proportional to the temperature. At higher temperatures, particles move more quickly (they have more energy) and there are more collisions in a given time.

More important, however, is the fact that particles will react only if, on collision, they have at least the minimum amount of energy, which is known as the activation energy.

> **Definition**
> The **activation energy** of a reaction is the minimum energy required for reaction to occur.

At higher temperatures, the mean energy of the particles is increased. The Maxwell–Boltzmann curves in Fig 4 show that, if the activation energy for a reaction is E_a, the number of molecules with energy greater than E_a (as shown by the shaded area) is much greater at temperature T_2 than at the lower temperature T_1.

Fig 4
Molecules with energy greater than E_a at different temperatures

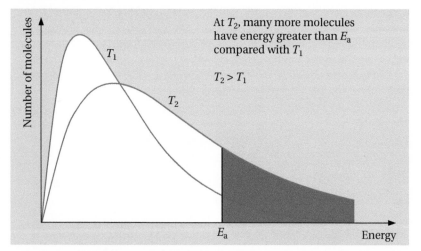

At T_2, many more molecules have energy greater than E_a compared with T_1

$T_2 > T_1$

The number of collisions between molecules with sufficient energy to react, i.e. the number of productive collisions, and therefore the rate of reaction, is very much greater at the higher temperature. Consequently, small temperature increases can lead to large increases in rate, as shown in Fig 5.

Many reactions – including several which are very **exothermic** – do not occur because the activation energy required is too high. For example, petrol reacts with oxygen in air in a very **exothermic reaction**, but a petrol–air mixture exists in the petrol tank of a car and will react only if sparked.

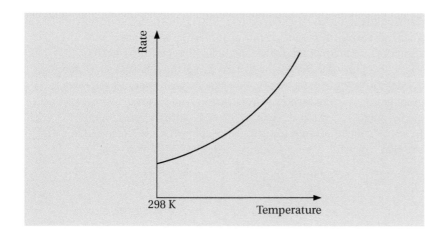

Fig 5
Change of rate as temperature rises

3.1.5.4 Effect of concentration, pressure and surface area on reaction rate

Concentration

Increasing the concentration of a reagent increases the number of particles in a given volume and so increases the collision rate, and hence the chance of productive collisions. This change increases the **rate of reaction** (if the reagent appears in the rate equation).

As a reaction proceeds, reagents are used up, so their concentrations fall. The rate is therefore at its greatest at the start of a reaction. On a concentration–time graph, the initial gradient is the steepest (most negative). The gradient falls to zero at the completion of the reaction, as shown in Fig 6.

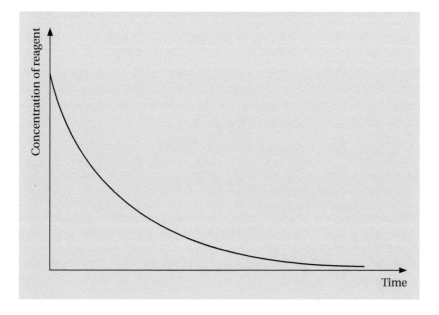

Fig 6
Fall in concentration of reagent with time at constant temperature

Pressure

Increasing the pressure of a gas increases the number of particles in a given volume and so produces the same effects as an increase in concentration.

Surface area

When one reagent is a solid, the rate of its reaction with a gas or with a substance in solution is increased if the solid is broken into smaller pieces. This process increases the surface area of the solid and allows more collisions to occur with particles of the other reagent. For example, when a given mass of calcium carbonate is reacted with an excess of hydrochloric acid and the volume of carbon dioxide produced is plotted against time, the initial gradient of the graph is much steeper when powdered carbonate is used (Fig 7, curve A) than when lumps are reacted (Fig 7, curve B).

Fig 7
Volume of CO_2 against time at constant temperature

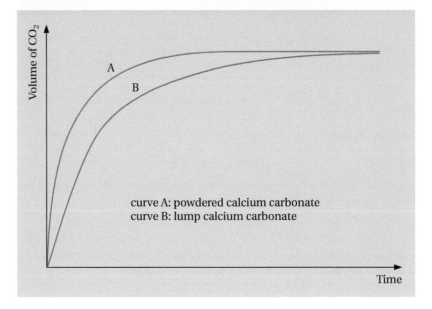

curve A: powdered calcium carbonate
curve B: lump calcium carbonate

Note that the same amount of calcium carbonate must have been used up in each experiment since the final volume of CO_2 is the same in both cases.

When an ionic solid is dissolved in a solvent, its particles are completely separated so that the rate is increased even further, and the reaction may become almost instantaneous. Precipitates form as soon as the correct solutions are mixed, since the free ions in solution can easily collide and react.

3.1.5.5 Catalysts

A **catalyst** is a substance which alters the rate of a reaction without itself being consumed during the reaction. Most of the catalysts used are positive catalysts: they increase the rate of reaction.

> **Definition**
>
> A **catalyst** alters the rate of a chemical reaction without itself being consumed.

A positive catalyst operates by providing an alternative route or reaction mechanism which has a lower activation energy E_{cat} than the uncatalysed route. Fig 8 shows a reaction profile for a catalysed and an uncatalysed reaction. Note that the catalyst has no effect on the overall enthalpy change for

Notes

An example of a negative catalyst, i.e. one which slows down a reaction, is antimony oxide, which is used as a flame retardant in plastics such as PVC.

the reaction. The catalyst also has no effect on the equilibrium position since this depends only on the relative energies of the reactants and products.

The reaction profile in Fig 8 shows a one-step catalysed reaction. In many cases the catalysed reaction occurs in more than one step and a double-humped reaction profile will be seen, as in Fig 9.

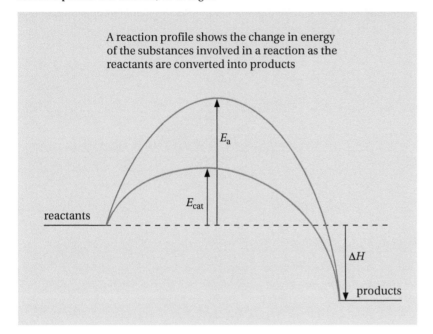

A reaction profile shows the change in energy of the substances involved in a reaction as the reactants are converted into products

Fig 8
Reaction profile showing an uncatalysed and a one-step catalysed reaction

Essential Notes

Catalytic action is covered in *Collins Student Support Materials: A-Level year 2 – Inorganic and Relevant Physical Chemistry*, section 3.2.5.6.

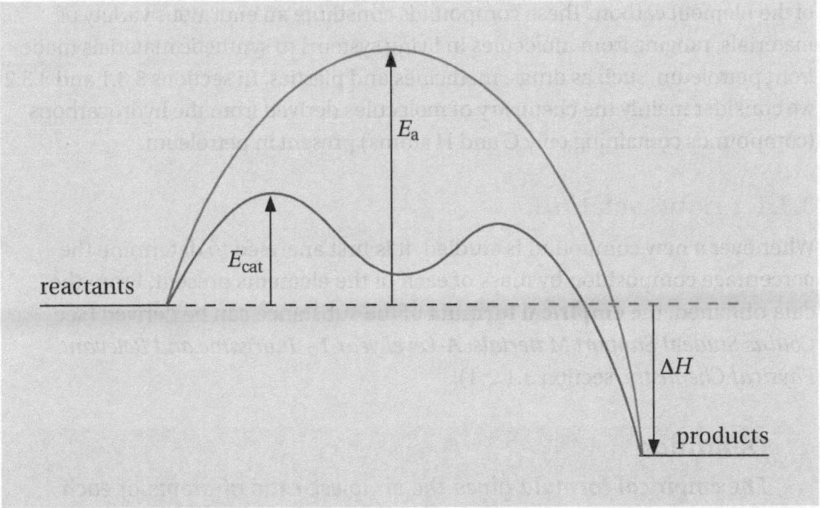

Fig 9
Reaction profile showing an uncatalysed and a two-step catalysed reaction

Figure 10 shows the effect of a catalyst on Maxwell–Boltzmann distribution. The lowering of activation energy caused by the catalyst increases the number of molecules with greater energy than the activation energy. This change is shown by there being a larger area to the right of the E_{cat} line, compared to the area to the right of the E_a line. Consequently, there is an increase in the number of productive collisions and therefore an increase in the rate of reaction.

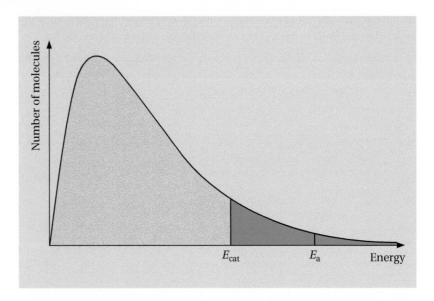

Fig 10
Effect of a catalyst on activation energies

3.3 Organic chemistry

3.3.1 Introduction to organic chemistry

Organic chemistry is the study of the many millions of covalent compounds of the element carbon. These compounds constitute an enormous variety of materials, ranging from molecules in living systems to synthetic materials made from petroleum, such as drugs, medicines and plastics. In sections 3.3.1 and 3.3.2 we consider mainly the chemistry of molecules derived from the hydrocarbons (compounds containing only C and H atoms) present in petroleum.

3.3.1.1 Nomenclature

Whenever a new compound is studied, it is first analysed to determine the percentage composition by mass of each of the elements present. From the data obtained, the **empirical formula** of the substance can be derived (see *Collins Student Support Materials: A-Level year 1 – Inorganic and Relevant Physical Chemistry*, section 3.1.2.4).

> **Definition**
> *The **empirical formula** gives the simplest ratio of atoms of each element in a compound.*

Although the empirical formula of a compound X, CH_2O, gives the simplest ratio of atoms, this ratio can be found in many different molecules. From the empirical formula we can determine the **molecular formula**.

Definition

*The **molecular formula** gives the actual number of atoms of each element present in a molecule.*

The molecular formula must be a multiple of the empirical formula, i.e. $(CH_2O)_n$ in this example. In order to discover the value of n in the formula $(CH_2O)_n$, we need to know the relative molecular mass of compound X. The mass of the empirical formula CH_2O is 30. Thus, if the relative molecular mass of X is found to be 60, then the value of n is 2 and the molecular formula is $(CH_2O)_2$, which is more usually written as $C_2H_4O_2$.

This molecular formula, $C_2H_4O_2$, can represent several compounds in which the two atoms of carbon, four atoms of hydrogen and two atoms of oxygen are arranged differently. These different molecules are called **isomers**.

The molecular formula for X, $C_2H_4O_2$, can represent several different compounds. The structures of these compounds can be shown by **displayed formulae**, **structural formulae** or **skeletal formulae**.

Definition

*A **displayed formula** shows all the bonds present in a molecule.*

Displayed formulae for two isomers of $C_2H_4O_2$ are shown in Fig 11.

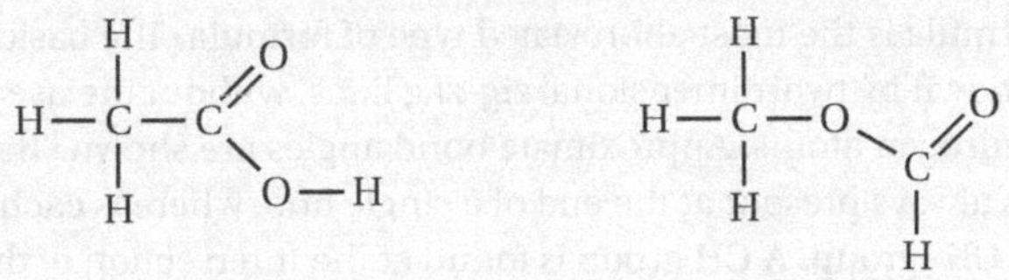

However, it is not necessary to show all of the bonds in a molecule to give a true picture of its structure, so it is simpler and more common to use a structural formula.

Definition

*A **structural formula** shows the unique arrangement of atoms in a particular molecule in a simplified form, without showing all the bonds.*

Structural formulae for two compounds with molecular formula $C_2H_4O_2$ are shown in Fig 12.

Essential Notes

Isomerism is considered in detail later in this book (section 3.3.1.3) and in *Collins Student Support Materials: A-Level year 2 – Organic and Relevant Physical Chemistry*, section 3.3.7.

Fig 11
Two structural isomers of $C_2H_4O_2$ shown as displayed formulae

Fig 12
Two structural isomers of $C_2H_4O_2$ shown as structural formulae

These structural formulae do not show all the bonds present, but make it clear which compound is which by displaying clearly which functional groups are present.

Representation of these two compounds either by displayed or structural formulae is an approximation since the molecules are three-dimensional, not planar as shown here. Displayed formulae can be cumbersome to draw so, particularly in equations, structural or simplified structural formulae are usually used, as in Equation 1.

$$CH_3COOH + CH_3CH_2OH \rightarrow CH_3COOCH_2CH_3 + H_2O \qquad \text{(Eqn 1)}$$

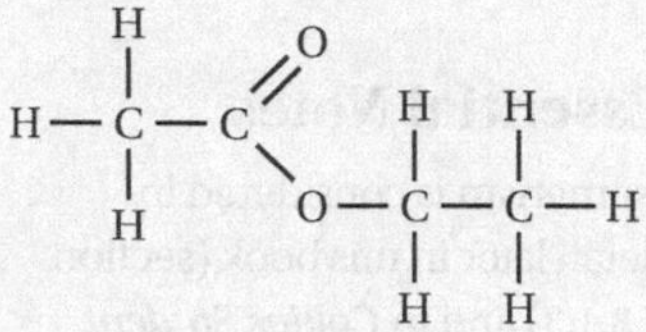

The organic product of the reaction in Equation 1 is more simply written as $CH_3COOCH_2CH_3$ rather than the displayed formula in Fig 13.

Molecular formulae, however, should not be used in equations as this could lead to confusion between structural isomers (compounds with the same molecular formula but different structures, see section 3.3.1.3). Equation 1 is unambiguous, whereas the compounds in Equation 2 could easily be misinterpreted.

$$C_2H_4O_2 + C_2H_6O \rightarrow C_4H_8O_2 + H_2O \qquad \text{(Eqn 2)}$$

To identify a particular compound from among the possible structural isomers, the chemical and/or physical properties of the compound need to be studied. Nowadays the traditional technique of elemental analysis has largely been superseded by instrumental methods such as infrared spectroscopy (see this book, section 3.3.6.3), mass spectrometry (see this book, section 3.3.6.2) and nuclear magnetic resonance spectroscopy (see *Collins Student Support Materials: A-Level year 2 – Organic and Relevant Physical Chemistry*, section 3.3.15).

A skeletal formula is the most abbreviated type of formula. The basic carbon skeleton is shown by two-dimensional zig-zag lines, without the use of any carbon or hydrogen atoms. Approximate bond angles are shown. Thus, a CH_3 group is always present at the end of a single line, whereas each 'elbow' represents a CH_2 group. A CH group is found at the intersection of three lines and a quaternary C atom is present at a four-line junction. All four possibilities are seen in the skeletal representation of 2,2,3-trimethylpentane. Functional groups can, of course, be attached to skeletal structures.

Definition

*A **skeletal formula** is the most abbreviated type of formula. The basic carbon skeleton is shown by two-dimensional zig-zag lines, without the use of any carbon or hydrogen atoms.*

Some examples are shown in Fig 14. Note that the cyclic structure often used for benzene (C_6H_6) is effectively skeletal.

The organic compounds in this unit are almost all based on the series of hydrocarbons called alkanes (see this book, section 3.3.2), in which one of the hydrogen atoms may be replaced by an atom or group of atoms called a **functional group**.

Fig 13
Displayed formula of $C_4H_8O_2$

Notes

Structures must satisfy the rules of:

- 4 bonds to each carbon (tetravalent)
- 2 bonds to each oxygen (divalent) and
- 1 bond to each hydrogen (monovalent).

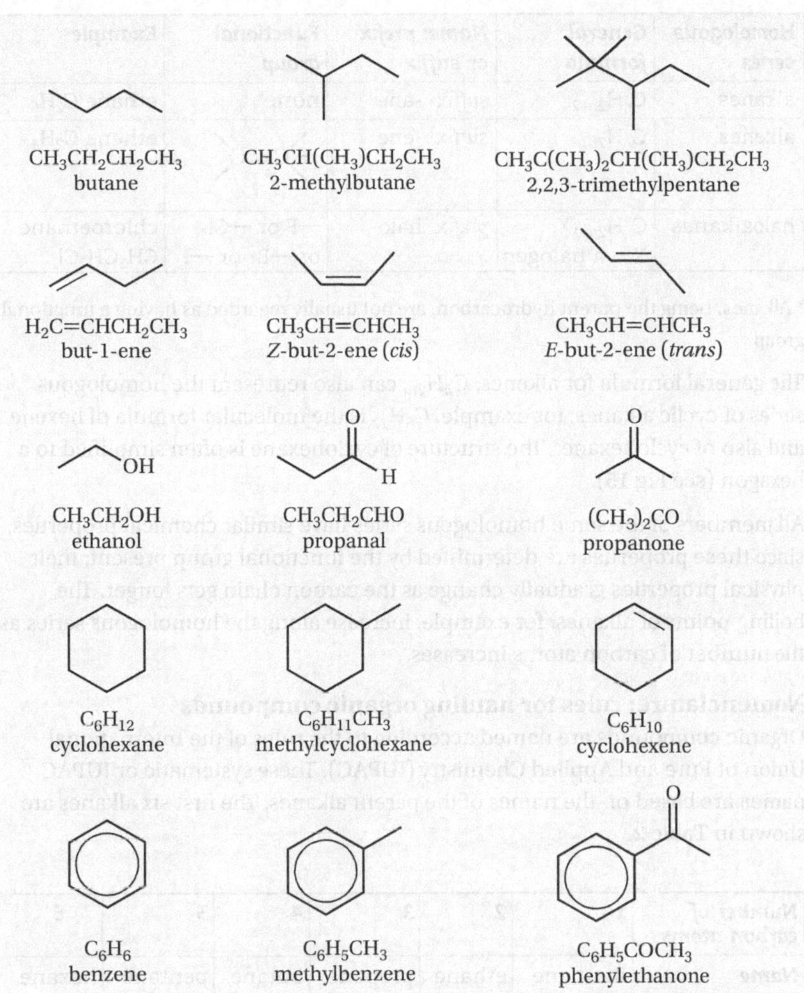

Fig 14
Some examples of skeletal formulae

$CH_3CH_2CH_2CH_3$
butane

$CH_3CH(CH_3)CH_2CH_3$
2-methylbutane

$CH_3C(CH_3)_2CH(CH_3)CH_2CH_3$
2,2,3-trimethylpentane

$H_2C=CHCH_2CH_3$
but-1-ene

$CH_3CH=CHCH_3$
Z-but-2-ene (cis)

$CH_3CH=CHCH_3$
E-but-2-ene (trans)

CH_3CH_2OH
ethanol

CH_3CH_2CHO
propanal

$(CH_3)_2CO$
propanone

C_6H_{12}
cyclohexane

$C_6H_{11}CH_3$
methylcyclohexane

C_6H_{10}
cyclohexene

C_6H_6
benzene

$C_6H_5CH_3$
methylbenzene

$C_6H_5COCH_3$
phenylethanone

Definition

A *functional group* is an atom or group of atoms which, when present in different molecules, causes them to have similar chemical properties

The functional group is the reactive part of a molecule; the properties of the molecule are largely determined by the nature of the functional group.

A family of molecules which all contain the same functional group, but an increasing number of carbon atoms, is called an **homologous series**, and can be represented by a general formula.

The general formulae and functional groups for the homologous series of alkanes, alkenes and haloalkanes are shown in Table 1.

15

Table 1
Homologous series, general formulae
and functional groups

Essential Notes

Prefix means added before the rest of the name; *suffix* means the ending of the name.

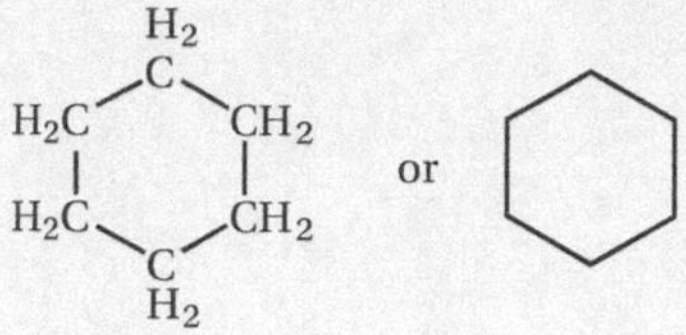

Fig 15
Two representations of cyclohexane

Notes

In the molecule $C_nH_{2n+1}X$, C_nH_{2n+1} is called an alkyl group ('ane' in the name is replaced by 'yl'). Thus, CH_3 is called methyl. Alkyl groups are represented by the letter R.

Essential Notes

Each successive molecule in a homologous series contains an additional $—CH_2—$ group.

Table 2
The first six alkanes

Notes

$$CH_3 — CH — CH_3$$
$$|$$
$$CH_3$$

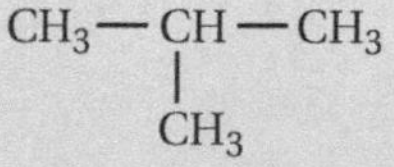

is called methylpropane as the longest chain has 3 carbon atoms (propane) with a one-carbon branch (methyl). No number is needed to show the postion of the methyl group as no other methylpropane is possible.

Table 1 Homologous series, general formulae and functional groups

Homologous series	General formula	Name: prefix or suffix	Functional group	Example
alkanes	C_nH_{2n+2}	*suffix* -ane	none*	ethane C_2H_6
alkenes	C_nH_{2n}	*suffix* -ene	$\diagdown C=C \diagup$	ethene C_2H_4
haloalkanes	$C_nH_{2n+1}X$ X = a halogen	*prefix* halo-	—F or —Cl or —Br or —I	chloroethane CH_3CH_2Cl

* Alkanes, being the parent hydrocarbon, are not usually regarded as having a functional group

The general formula for alkenes, C_nH_{2n}, can also represent the homologous series of cyclic alkanes; for example, C_6H_{12} is the molecular formula of hexene and also of cyclohexane. The structure of cyclohexane is often simplified to a hexagon (see Fig 15).

All members of the same homologous series have similar chemical properties, since these properties are determined by the functional group present; their physical properties gradually change as the carbon chain gets longer. The boiling points of alkanes, for example, increase along the homologous series as the number of carbon atoms increases.

Nomenclature: rules for naming organic compounds

Organic compounds are named according to the rules of the International Union of Pure and Applied Chemistry (IUPAC). These systematic or IUPAC names are based on the names of the parent alkanes. The first six alkanes are shown in Table 2.

Number of carbon atoms	1	2	3	4	5	6
Name	methane	ethane	propane	butane	pentane	hexane
Formula	CH_4	C_2H_6	C_3H_8	C_4H_{10}	C_5H_{12}	C_6H_{14}

To assign the name of an alkane derivative, first look for the longest carbon chain in the skeleton; the number of carbons in this chain determines the stem of the name. Thus, if there are two carbon atoms in the longest chain, the stem name will be ethan-; if there are five carbon atoms, the stem name will be pentan-.

In many compounds the carbon skeleton is branched. The names of the side chains also depend on the number of carbon atoms in them, so that:

- a one-carbon branch is called methyl (CH_3-)
- a two-carbon branch is called ethyl (CH_3CH_2-)
- a three-carbon branch is called propyl ($CH_3CH_2CH_2-$)

The position of any branch on the chain must also be made clear. This is achieved by numbering the carbon atoms in the skeleton so as to keep the numbers used as low as possible when indicating the position of any branches. For example, the molecule in Fig 16 is called 2-methylpentane (numbering from the right) and not 4-methylpentane (numbering from the left).

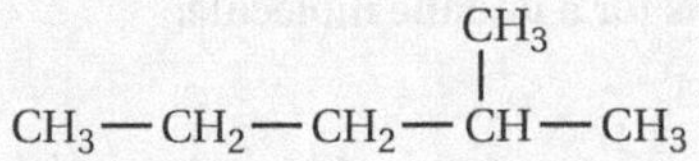

$$CH_3-CH_2-CH_2-\underset{\underset{\displaystyle CH_3}{|}}{CH}-CH_3$$

Fig 16
2-methylpentane

Note that, in these examples, the molecules are really three-dimensional with each carbon in the alkyl groups surrounded **tetrahedrally** by four bonds, with bond angles of approximately 109.5° (see Fig 17).

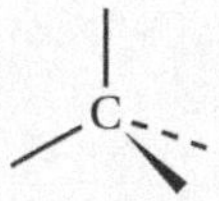

Fig 17
The tetrahedral arrangement of bonds around carbon

However, on paper it is usually simpler to represent a structure as if it had bond angles of 90° or 180°. For example, the three structures in Fig 18 all represent 3-methylpentane.

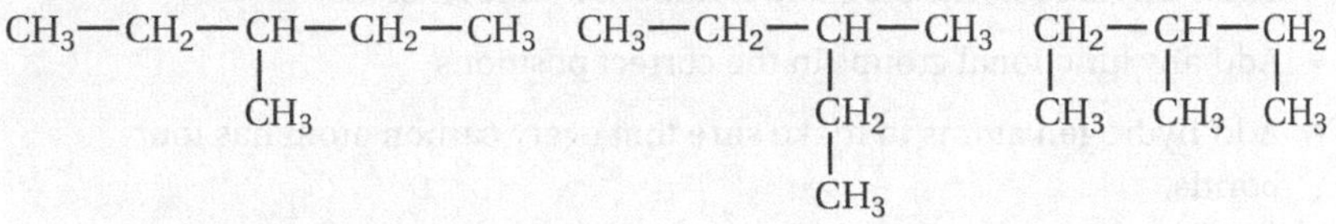

$$CH_3-CH_2-\underset{\underset{\displaystyle CH_3}{|}}{CH}-CH_2-CH_3 \quad CH_3-CH_2-\underset{\underset{\underset{\displaystyle CH_3}{|}}{\underset{\displaystyle CH_2}{|}}}{CH}-CH_3 \quad \underset{\underset{\displaystyle CH_3}{|}}{CH_2}-\underset{\underset{\displaystyle CH_3}{|}}{CH}-\underset{\underset{\displaystyle CH_3}{|}}{CH_2}$$

Fig 18
Different representations of 3-methylpentane

Naming molecules containing functional groups

- The *type* of functional group present is indicated by either a prefix or a suffix on the alkane stem (see Table 1).

- The *position* of the functional group is usually indicated by a number; e.g. 2-chloropropane (see Fig 19) has the chlorine atom on the second (middle) carbon atom.

- When two or more of a specific functional group are present, the *number* of substituents is shown by using the multipliers *di* for two, *tri* for three or *tetra* for four; for example dichloromethane, CH_2Cl_2, and tetrachloromethane, CCl_4.

$$CH_3-\underset{\underset{\displaystyle Cl}{|}}{CH}-CH_3$$

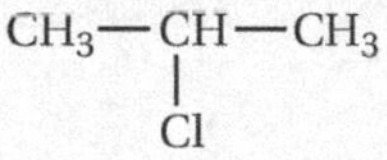

Fig 19
The structure of 2-chloropropane

- Numbers 1,2, etc. must also be used to show the position of each functional group. Commas are used between numbers and hyphens between numbers and letters; for example:

  CH_3CCl_3 is called 1,1,1-trichloroethane

  $CH_2ClCHCl_2$ is called 1,1,2-trichloroethane

- If more than one type of functional group is present, the positions and names are listed as prefixes in alphabetical order, for example:

  $CH_3CHBrCH_2Cl$ is 2-bromo-1-chloropropane

- Multipliers are ignored when ordering substituents alphabetically; tribromo- will always come before dichloro-

- The suffix -ene for alkenes can be placed in front of other suffixes and is shortened to en if followed by a number, for example:

  $H_2C=CH-CH_2OH$ is prop-2-en-1-ol

> **Notes**
>
> When naming compounds, look for:
> - the longest carbon chain
> - functional group(s)
> - the number of substituents
> - where they are.

Trivial names

Where systematic names become complicated, trivial names are often used; for example, lindane is one of the three-dimensional isomers of 1,2,3,4,5,6-hexachlorocyclohexane.

Lindane is used as an insecticide in agriculture and also to treat lice.

Alternative representations for a lindane molecule:

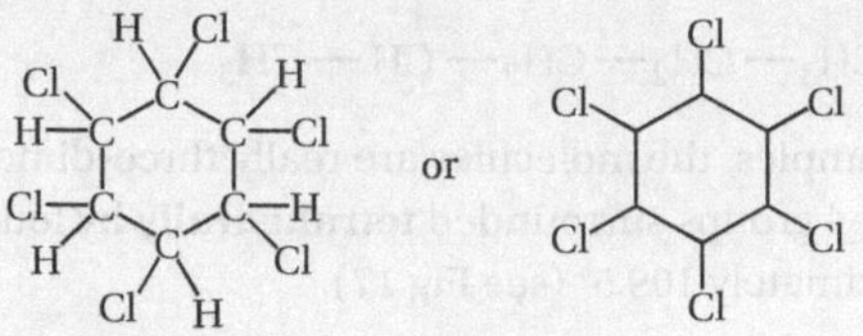

Drawing a structure from a given name

- Use the name to identify the number of carbon atoms in the longest chain

- Draw this carbon skeleton and number the carbon atoms

- Add any functional groups in the correct positions

- Add hydrogen atoms to make sure that every carbon atom has four bonds.

Example

The structure of 3-methylpent-2-ene can be deduced as follows:

pent- skeleton $\quad\quad$ C—C—C—C—C

pent-2-ene skeleton $\quad$ $\overset{1}{C}-\overset{2}{C}=\overset{3}{C}-\overset{4}{C}-\overset{5}{C}$

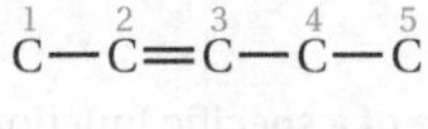

3-methylpent-2-ene skeleton $\quad$ $C-C=\overset{\overset{\displaystyle CH_3}{|}}{C}-C-C$

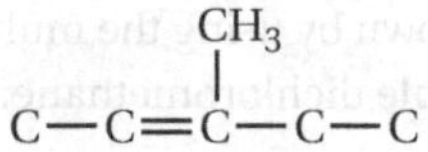

3-methylpent-2-ene complete $\quad$ $H-\overset{\overset{\displaystyle H}{|}}{\underset{\underset{\displaystyle H}{|}}{C}}-\overset{\overset{\displaystyle H}{|}}{C}=\overset{\overset{\displaystyle CH_3}{|}}{C}-\overset{\overset{\displaystyle H}{|}}{\underset{\underset{\displaystyle H}{|}}{C}}-\overset{\overset{\displaystyle H}{|}}{\underset{\underset{\displaystyle H}{|}}{C}}-H$

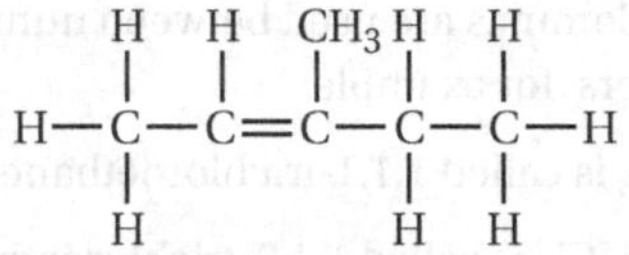

which can be written more simply as: $CH_3CH=C(CH_3)CH_2CH_3$

3.3.1.2 Reaction mechanisms

Reaction mechanisms seek to illustrate how a reaction takes place. Since the breaking of bonds and the formation of new bonds involve electrons, reaction mechanisms focus on the movement of electrons at various stages of a reaction sequence.

Instead of showing a reaction using a single equation, a reaction mechanism sets out the stages in a reaction, and has added notation to show how electrons move during these stages.

Since compounds in the same homologous series have similar chemical properties, it follows that their reactions occur by similar mechanisms. For example, substitution reactions of alkanes occur via a free-radical mechanism (see below).

Free-radical mechanisms

A free radical is an atom or molecule with an unpaired electron. We place a dot (•) alongside the species to show the unpaired electron.

For example, the uv radiation in sunlight can cause a chlorine molecule to break its **covalent bond**, producing two free radicals:

$$Cl_2 \rightarrow Cl\bullet + Cl\bullet$$

Free radicals are highly reactive so, once formed, they trigger a rapid sequence of further radical reactions (see section 3.3.2.4).

In some texts, single-barbed curly arrows are used to annotate radical mechanisms, but this is not required for this specification.

Other mechanisms

When free radicals are not involved in a reaction mechanism, the movement of an electron pair is shown by the use of a curly arrow with two barbs.

The two barbs specifically indicate the movement of two electrons.

The formation of a covalent bond is shown by a curly arrow which starts from a lone electron pair or from another covalent bond. The breaking of a covalent bond is shown by a curly arrow which starts from the bond.

These uses of curly arrows are shown in Fig 20.

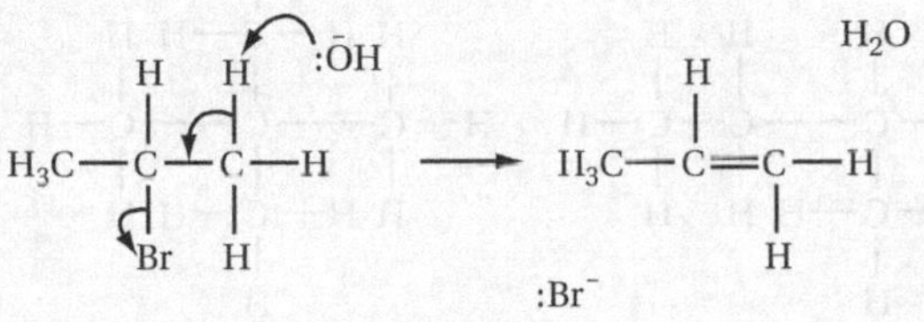

Fig 20
Different uses of curly arrows

For details of mechanisms included in this book, see sections 3.3.2.4, 3.3.3.1, 3.3.3.2, 3.3.4.2 and 3.3.5.3.

3.3.1.3 Isomerism

Isomerism occurs where molecules with the same molecular formula have their atoms arranged in different ways. Isomerism is divided into two main types, which are also themselves subdivided:

- **structural isomerism**
- **stereoisomerism**

Structural isomerism

> **Definition**
> **Structural isomers** *are compounds with the same molecular formula, but with different structures.*

The different structures can arise in any of three different ways:

- chain isomerism
- position isomerism
- functional group isomerism.

Chain isomerism

Chain isomers occur when there are two or more ways of arranging the carbon skeleton of a molecule. For example, C_4H_{10} can be butane or 2-methylpropane, as shown in Fig 21.

Fig 21
The isomers of C_4H_{10}

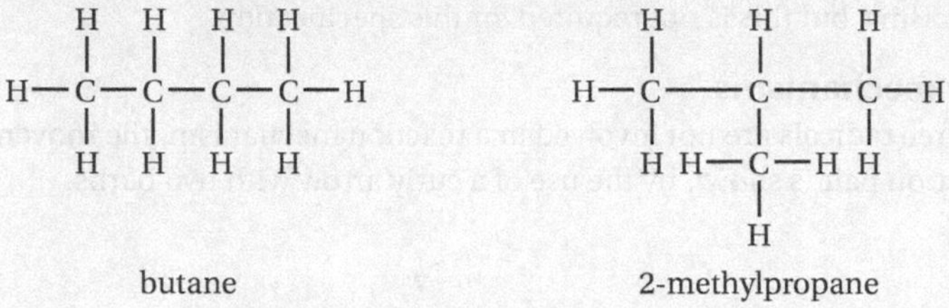

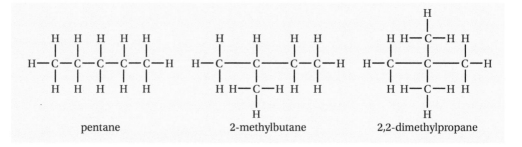

butane   2-methylpropane

Essential Notes

The number 2 is optional in 2-methylpropane, as was explained earlier (see this book, section 3.1.1.1).

The three isomers of C_5H_{12} are pentane, 2-methylbutane and 2,2-dimethylpropane, as shown in Fig 22.

Fig 22
The isomers of C_5H_{12}

pentane   2-methylbutane   2,2-dimethylpropane

These isomers have similar chemical properties, but slightly different physical properties. Branched isomers have smaller volumes, **weaker van der Waals' forces** (see *Collins Student Support Materials: A-Level year 1 – Inorganic and Relevant Physical Chemistry*, section 3.1.3.7) and therefore lower boiling points.

The number of structural isomers of alkanes rises steeply as the number of carbon atoms increases, as shown in Table 3.

Position isomerism

Position isomers have the same carbon skeleton and the same functional group, but the functional group is joined at different places on the carbon skeleton. For example:

$CH_3CH_2CH_2Br$ $CH_3CHBrCH_3$
1-bromopropane   2-bromopropane

$H_2C{=}CHCH_2CH_3$ $CH_3CH{=}CHCH_3$
but-1-ene   but-2-ene

Again, such isomers have similar chemistry because they have the same functional group, but the different positions can cause some differences in properties.

Number of carbon atoms	Number of isomers
1	1
2	1
3	1
4	2
5	3
6	5
7	9
8	18
9	35
10	75
11	159
12	355
13	802
14	1,858
15	4,347
20	366,319
25	36,797,588
30	4,111,846,763
40	62,491,178,805,831

Table 3
The number of structural isomers of some alkanes

Functional group isomerism

Functional group isomers contain different functional groups and therefore have different chemical properties. For example, the molecular formula C_6H_{12} applies to cyclic alkanes, such as cyclohexane, and also to alkanes, such as hex-l-ene:

cyclohexane   hex-l-ene

The molecular formula C_3H_6O represents the aldehyde propanal and also the ketone propanone:

CH_3CH_2CHO CH_3COCH_3
propanal   propanone

Notes
These pairs of isomers can be distinguished by means of a simple chemical test (see this book, section 3.3.6.1).

Stereoisomerism

The two types of stereoisomerism are **E–Z stereoisomerism** and optical isomerism (see *Collins Student Support Materials: A-Level year 2 – Organic and Relevant Physical Chemistry*, section 3.3.7).

Definition
Stereoisomers are compounds which have the same structural formula but the bonds are arranged differently in space.

E–Z stereoisomerism

Because of restricted rotation at the C=C bond, Z or cis and E or trans forms occur when there is suitable substitution:

Z or cis   E or trans

Essential Notes

E stands for entgegen (German for opposite). Z stands for zusammen (German for together). E–Z stereoisomerism is also known as geometrical or cis–trans isomerism.

It is not possible to have E–Z stereoisomerism when there are two identical groups joined to the same carbon atom in a double bond. Restricted rotation about a C=C bond arises due to the interaction between the two adjacent p-orbitals of the carbon atoms, forming a π-bond. Disruption of the π-bond requires significantly more energy than is available at room temperature, so that rotation does not occur readily.

E means that two groups, which may be identical, are on opposite sides of the double bond and Z means they are on the same side. For example, but-2-ene exists as two forms that differ only in the arrangement of the bonds in space (see Fig 23).

Fig 23
The two E–Z stereoisomers of but-2-ene

Z-but-2-ene (cis)   E-but-2-ene (trans)

Note that methylpropene is a structural isomer, but not an E–Z stereoisomer of but-2-ene, because it is not possible to have E–Z stereoisomerism when there are two identical groups joined to the same carbon atom in a double bond (see Fig 24).

Fig 24
Structure of methylpropene

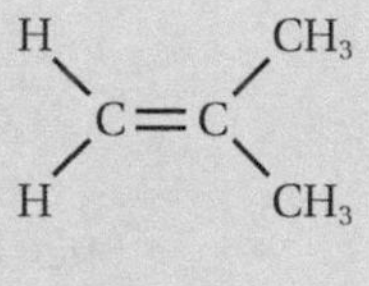

methylpropene

Methylpropene has two methyl groups on one carbon and two hydrogen atoms on the other. It is a structural isomer of but-2-ene, as also are but-1-ene and the cyclic alkanes cyclobutane and methylcyclopropane (see Fig 25).

Fig 25
Other isomers of methylpropene

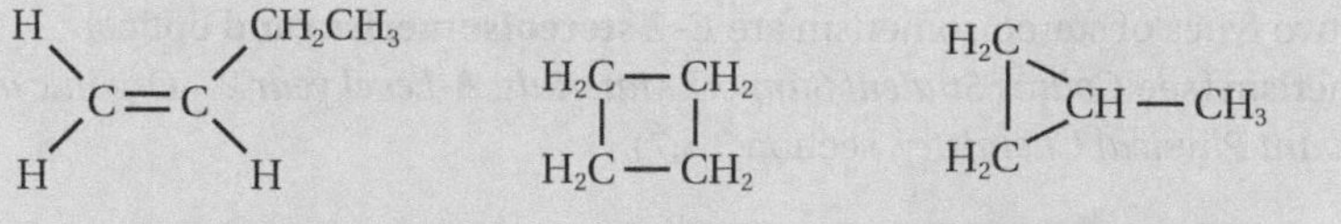

but-1-ene   cyclobutane   methylcyclopropane

Cahn–Ingold–Prelog (CIP) priority rules

The *E–Z* system, as outlined here, works well in simple cases. However, when three or four different groups are attached to the C=C bond, a more detailed approach is called for.

So, to deal with such cases, and to enable *E–Z* labelling of all alkenes, the *E–Z* system is extended using Cahn–Ingold–Prelog (CIP) rules. These rules assign the priorities for use when naming such compounds.

Looking at the two atoms attached directly to the left-hand carbon of the double bond, priority is given to the atom with the highest atomic number. Similarly, for the right-hand carbon of the double bond, the atom with the highest atomic number is assigned the highest priority.

An example is shown in Fig 26:

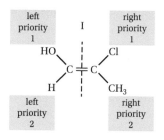 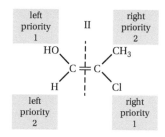

Notes

E–Z stereoisomers have different physical and chemical properties. Z-isomers usually have slightly higher boiling points as they will have some **polarity**, whereas E-isomers are less polar. The boiling point of Z-but-2-ene is 4 °C whereas the boiling point of E-but-2-ene is 1 °C. E-isomers, however, often have higher melting points because they pack together better. For example, the melting point of Z-but-2-ene is −139 °C whereas the melting point of E-but-2-ene is −106 °C.

Fig 26
Assessing priority using the CIP rules

For the left-hand double-bonded carbon of both structures, priority is as shown: i.e. OH > H, since O has a higher atomic number than H. For the right-hand double-bonded carbon of both structures, priority is as shown: i.e. $Cl > CH_3$, since Cl has a higher atomic number than C.

It can now be seen that:

- In structure I, the two highest-priority atoms are the same side of the double bond (above it), so this is named as the *Z*-isomer.

- In structure II, the two highest-priority atoms are the opposite sides of the double bond (above and below it), so this is named as the *E*-isomer.

When priority has to be decided between two groups which have the same first atom attached to the double-bonded carbon, it is necessary to look at the second atoms along the two groups.

For example, to decide priority between CH_3 and CH_2CH_3, we must look at the atoms next along from the first carbons. With CH_3, H is next. With CH_2CH_3, we have a second C (and its attached H atoms). Since C has a higher atomic number than H, then CH_2CH_3 takes priority over CH_3.

Similarly, CH_2OH would take priority over CH_2CH_3, as the O has higher atomic number than the second C of CH_2CH_3.

3.3.2 Alkanes

Essential Notes

Saturated compounds contain only single bonds; unsaturated compounds contain one or more double or triple bonds.

Notes

Electrophiles are electron-pair acceptors and seek electron-rich sites (see this book, section 3.3.4.2).

Nucleophiles are electron-pair donors and seek electron-deficient sites (see this book, section 3.3.3.1).

Fig 27
Fractionating column used to distill crude oil

3.3.2.1 Fractional distillation of crude oil

Petroleum

Petroleum or crude oil is a complex mixture of hydrocarbons, mainly alkanes; it is derived from the remains of sea creatures and plants which sank to the bottom of the oceans millions of years ago. Subsequent deposits compressed this material and the high pressures and temperatures which developed – and also the absence of air – converted it into oil and gas.

Alkanes

Alkanes are the homologous series of **saturated hydrocarbons** with the general formula C_nH_{2n+2}. The first six alkanes are listed in Table 2. The lower alkanes are gases at room temperature; their boiling points increase with the number of carbon atoms because the strength of van der Waals' forces between the molecules increases. This increase in boiling points allows crude oil to be separated by fractional distillation (see below).

Alkanes contain only carbon–carbon and carbon–hydrogen bonds; these bonds are relatively strong and are non-polar. Consequently, alkanes are unreactive towards acids, alkalis, electrophiles and nucleophiles. In common with all hydrocarbons, however, they burn in air or oxygen with highly exothermic reactions; hence, they are important for use as fuels (see this book, section 3.3.2.3).

Fractional distillation

The complex mixture of hydrocarbons in crude oil, mainly alkanes, is separated into less complicated mixtures, or fractions, by fractional distillation (primary distillation – so called as it is the first stage in the separation process).

A fractionating column or tower is the name given to the long vertical tube used in fractional distillation (see Fig 27). The crude oil is heated and the vapour/liquid mixture passed into the tower. The top of the tower is cooler than the

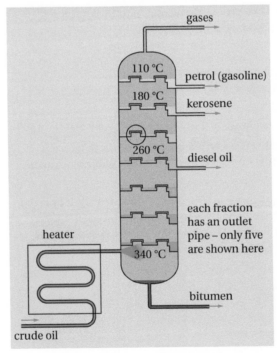

bottom. The temperature gradient in the tower allows separation of the petroleum mixture into fractions depending on the boiling points of the hydrocarbons present. Only the most volatile components, those with low boiling points, reach the top; others condense in trays at different levels up the tower and are drawn off.

The residue from primary distillation still contains useful materials, such as lubricating oil and waxes; these boil above 350 °C at atmospheric pressure. At such high temperatures, some of the components in the residue decompose. To avoid this, the residue is further distilled under reduced pressure (vacuum distillation). Using this method, the remaining hydrocarbons can be distilled at a lower temperature where decompostion does not occur.

The major fractions and their uses are shown in Table 4.

Name of fraction	Boiling range/°C (approx)	Uses	Length of carbon chain (approx)
LPG (liquefied petroleum gas)	up to 25	Calor Gas, Camping Gaz	1–4
petrol (gasoline)	40–100	petrol	4–12
naphtha	100–150	petrochemicals	7–14
kerosene (paraffin)	150–250	jet fuel, petrochemicals	11–15
gas oil (diesel)	220–350	central heating fuel, petrochemicals	15–19
mineral oil (lubricating oil)	over 350	lubricating oil, petrochemicals	20–30
fuel oil	over 400	fuel for ships and power stations	30–40
wax, grease	over 400	candles, grease for bearings, polish	40–50
bitumen	over 400	roofing, road surfacing	above 50

Table 4
Fractions from crude oil

The composition of crude oil varies from one place to another. In general, however, the amount of each fraction produced by distillation does not match the demand (see Table 5).

| Fraction | Approximate % | |
	Crude oil	Demand
gases	2	4
petrol and naphtha	16	27
kerosene	13	8
gas oil	19	23
fuel oil and bitumen	50	38

Table 5
Supply and demand for oil fractions

A higher proportion of the high-value products (such as petrol) is used commercially than occurs naturally, while there is not enough demand for some of the heavier fractions. To solve this imbalance, larger alkane molecules are broken up into smaller molecules in a process called **cracking**.

Notes
Fractional distillation is a *physical* process. Energy is needed only to separate molecules from each other, that is to overcome the van der Waals' forces.

Notes
Cracking is a *chemical* process. Energy is needed to break C—C bonds.

3.3.2.2 Modification of alkanes by cracking

Cracking

Hydrocarbon cracking involves breaking carbon–carbon and carbon–hydrogen bonds. Two main processes are used: **thermal cracking** and **catalytic cracking**. In general, large alkanes are cracked to form smaller alkanes, alkenes and sometimes also hydrogen:

high M_r alkanes → smaller M_r alkanes + alkenes (+ hydrogen)

These smaller alkanes are more in demand by the petrochemical industry and are therefore termed 'higher-value products'.

Ethene is another valuable product of cracking and can be used to make the plastic poly(ethene), commonly called *polythene*.

When cracked, molecules may break up in several different ways to form a mixture of products which can be separated by fractional distillation. For example, two possible fragmentations of the $C_{14}H_{30}$ molecule are:

$$C_{14}H_{30} \rightarrow C_7H_{16} + C_3H_6 + 2C_2H_4$$
$$C_{14}H_{30} \rightarrow C_{12}H_{24} + C_2H_4 + H_2$$

Thermal cracking

Thermal cracking results in the formation of a high proportion of alkenes. For example:

$$C_{14}H_{30} \rightarrow C_8H_{18} + 3C_2H_4$$

The energy required for bond breaking is provided by heat; the temperatures employed range from 400 °C to 900 °C at pressures of up to 7000 kPa. At the lower end of this temperature range, carbon chains break preferentially towards the centre of the carbon chain of the molecule. With increasing temperature, the cracking shifts towards the end of the chain, leading to a greater percentage of low M_r alkenes. In order to avoid decomposition into the constituent elements, the length of exposure to high temperatures (residence time) has to be short, of the order of one second.

Catalytic cracking

Catalytic cracking involves the use of zeolite catalysts (crystalline aluminosilicates), at a pressure slightly above atmospheric and a temperature of about 450 °C. By this means, large alkanes are converted mainly into branched alkanes, cycloalkanes and aromatic hydrocarbons. For example:

linear (unbranched) alkane   branched-chain isomer of octane   cyclohexane

The proportion of alkenes produced is small, so that catalytic cracking is primarily used for producing motor fuels. Branched-chain alkanes burn more smoothly than unbranched chains. In an engine, because of the pressures involved, the fuel–air mixture may ignite before the spark is produced, causing 'knocking'. This problem is prevented by using branched-chain alkanes.

3.3.2.3 Combustion of alkanes

In common with all hydrocarbons, alkanes burn in air or oxygen in very exothermic reactions and so are used as fuels. In the presence of a plentiful supply of oxygen, *complete combustion* of alkanes occurs to form carbon dioxide and water. For example:

$$CH_4 + 2O_2 \rightarrow CO_2 + 2H_2O \qquad \Delta H^{\ominus} = -890 \text{ kJ mol}^{-1}$$

$$C_4H_{10} + 6\tfrac{1}{2}O_2 \rightarrow 4CO_2 + 5H_2O \qquad \Delta H^{\ominus} = -2880 \text{ kJ mol}^{-1}$$

As the number of carbon atoms increases, more oxygen is required per mole of hydrocarbon for complete combustion, and more energy is released.

When insufficient oxygen is available, *incomplete combustion* occurs. Water is formed together with carbon monoxide or carbon. For example, if a bunsen burner is used with the air-hole closed, the flame is not blue but yellow and luminous, because of the carbon particles it contains. Any apparatus heated in a luminous flame becomes coated in black soot:

$$CH_4 + O_2 \rightarrow C + 2H_2O$$

Incomplete combustion forming carbon monoxide is, however, much more of a hazard. Badly maintained gas central heating boilers may produce carbon monoxide because of an inadequate supply of air, and can cause accidental death by carbon monoxide poisoning:

$$CH_4 + 1\tfrac{1}{2}O_2 \rightarrow CO + 2H_2O$$

Internal combustion engines

Carbon monoxide is also formed by the incomplete combustion of petrol vapour in a car engine:

$$C_8H_{18} + 8\tfrac{1}{2}O_2 \rightarrow 8CO + 9H_2O$$

Motor-car engines also produce other pollutants, notably oxides of nitrogen and unburned hydrocarbons. The high temperature produced when the fuel is burnt provides sufficient activation energy for nitrogen to react with oxygen to form nitrogen monoxide:

$$N_2 + O_2 \rightarrow 2NO$$

On cooling, nitrogen monoxide reacts easily with more oxygen to form nitrogen dioxide. With water and more oxygen, nitric acid is formed, which can lead to **acid rain**:

$$2NO + O_2 \rightarrow 2NO_2$$

$$4NO_2 + 2H_2O + O_2 \rightarrow 4HNO_3$$

Nitrogen dioxide also reacts with oxygen or hydrocarbons in the presence of sunlight to form an irritating photochemical smog.

Notes
Complete combustion of hydrocarbons produces carbon dioxide and water.

Notes
Incomplete combustion of hydrocarbons produces water and carbon or carbon monoxide.

Notes
Cars with petrol engines produce carbon monoxide; diesel engines produce carbon and very small amounts of carbon monoxide.

Catalytic converters

These devices help to remove carbon monoxide, nitrogen oxides and hydrocarbons from car exhausts (see Fig 28). Converters contain a honeycomb of ceramic material onto which metals such as platinum, palladium and rhodium are spread in a thin layer. These metals catalyse reactions between the pollutants and help to remove up to 90% of the harmful gases. For example:

$$2CO + 2NO \rightarrow 2CO_2 + N_2$$

$$C_8H_{18} + 25NO \rightarrow 8CO_2 + 12\frac{1}{2}N_2 + 9H_2O$$

Overall, the pollutant gases – CO and NO_x and hydrocarbons – are replaced by CO_2, N_2 and H_2O, which are harmless.

Fig 28
The action of a catalytic converter

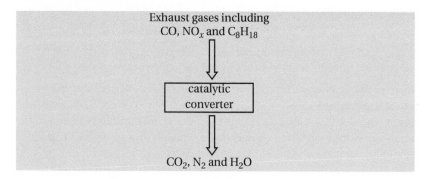

Exhaust gases including
CO, NO_x and C_8H_{18}

catalytic
converter

CO_2, N_2 and H_2O

Combustion of sulfur-containing impurities

The alkanes and other hydrocarbons in petroleum fractions usually occur together with sulfur-containing impurities. When these hydrocarbons are burned, the impurities are also oxidised and form sulfur dioxide; for example:

$$CH_3SH + 3O_2 \rightarrow CO_2 + 2H_2O + SO_2$$

Sulfur dioxide is a toxic gas and, being soluble in water, can cause **acid rain** by forming a solution of sulfurous acid, H_2SO_3. High in the atmosphere, ultraviolet radiation provides the energy for sulfur dioxide to react with oxygen to form sulfur trioxide. Sulfur trioxide is very soluble in water and forms sulfuric acid, which also occurs in acid rain.

A lot of sulfur dioxide is produced by the burning of fuels in power stations. However, this pollutant is not released into the atmosphere, but is removed from the gases passed up the chimney (flue) by a process called flue-gas desulfurisation.

Several alkaline substances can be used to remove the acidic sulfur dioxide. Some methods use calcium oxide (quicklime) which is easily obtained by heating calcium carbonate (limestone). The product of the reaction of calcium oxide and sulfur dioxide is calcium sulfite, $CaSO_3$.

$$CaO + SO_2 \rightarrow CaSO_3$$

Calcium sulfite is easily oxidised and forms hydrated calcium sulfate (gypsum), $CaSO_4 \cdot 2H_2O$, which is used to make plasterboard for the building industry.

Greenhouse gases

Combustion of all fossil fuels, including alkanes, eventually produces carbon dioxide. This gas, together with water vapour, methane and ozone (the

Essential Notes

As an alternative, calcium carbonate can be reacted directly to remove sulfur dioxide:

$$CaCO_3 + SO_2 \rightarrow CaSO_3 + CO_2$$

so-called **greenhouse gases**), is thought to contribute to global warming by absorbing infrared radiation.

Ultraviolet and visible radiation from the sun is absorbed by the Earth and emitted at much longer wavelengths as infrared radiation. This radiation is absorbed by some molecules in the atmosphere which trap the energy and prevent its escape. Not all gases absorb infrared radiation, but those which do are called greenhouse gases.

Many governments have introduced measures to control or reduce the emision of carbon dioxide. Car manufacturers, for instance, must publish data about CO_2 emissions from their vehicles. Other fuels or methods of producing energy are also being considered as alternatives to the combustion of fossil fuels.

3.3.2.4 Chlorination of alkanes

Alkanes such as methane do not react with chlorine at room temperature or in the dark. In the presence of ultraviolet light, however, a mixture of methane and chlorine will react at room temperature, forming hydrogen chloride and a mixture of chlorinated methanes. The mechanism for this process is called **free-radical substitution**.

A free radical (or radical for short) is a species which contains an odd number of electrons with one electron not paired with any other. Radicals can be single atoms or groups of atoms. A radical is represented by writing a dot (as in Cl•) to indicate the unpaired electron. Radicals are formed when a covalent bond breaks with an equal splitting of the bonding pair of electrons (**homolytic fission**).

The free-radical substitution **mechanism** occurs in several steps.

Initiation step

$$Cl_2 \rightarrow 2Cl\bullet$$

The ultraviolet light provides the energy needed to start the reaction by splitting some chlorine molecules into atoms (radicals). This process occurs first because the Cl—Cl bond in chlorine is weaker than the C—H bond in methane.

Propagation steps

$$Cl\bullet + CH_4 \rightarrow \bullet CH_3 + HCl$$

$$\bullet CH_3 + Cl_2 \rightarrow CH_3Cl + Cl\bullet$$

In each step, a radical is used and a new radical is formed, so that the process continues and leads to a **chain reaction**. Each step is exothermic, so that the chain reaction might produce an explosion. The overall reaction, which is the sum of the two propagation steps, can be represented by the equation:

$$CH_4 + Cl_2 \rightarrow CH_3Cl + HCl$$

Termination steps

When two radicals combine, they form a stable molecule and the sequence of reactions stops; the unpaired electrons in the radicals pair up to form a covalent bond. Two possible termination steps are:

$$Cl\bullet + \bullet CH_3 \rightarrow CH_3Cl$$

$$\bullet CH_3 + \bullet CH_3 \rightarrow CH_3CH_3$$

Essential Notes

The causes of global warming are not fully understood, the level of carbon dioxide in the atmosphere being only one of several factors involved.

Essential Notes

The absorption of infrared radiation is considered in more detail in this book, section 3.3.6.3.

Notes

Ultraviolet light consists of very high-energy radiation, enough to break the Cl—Cl bond

Notes

Chlorination of propane produces two monochloropropanes, 1-chloropropane and 2-chloropropane, in an approximate ratio of 3:1 because of the ratio of 6 CH_3 hydrogens to 2 CH_2 hydrogens in propane.

Such termination steps can lead to trace amounts of impurities, such as ethane, in the final product.

The termination step:

$$Cl\bullet + Cl\bullet \rightarrow Cl_2$$

is possible, but unlikely because the chlorine radicals collide with enough energy to separate again immediately.

Further substitution

The reaction of a chlorine radical with methane extracts a hydrogen radical (atom) to form HCl, as shown in the first propagation step. Chloromethane still contains three hydrogen atoms, and so further pairs of propagation steps are possible, leading to dichloromethane (CH_2Cl_2), trichloromethane ($CHCl_3$) and finally to tetrachloromethane (CCl_4). The propagation steps to form CH_2Cl_2 are:

$$Cl\bullet + CH_3Cl \rightarrow \bullet CH_2Cl + HCl$$
$$\bullet CH_2Cl + Cl_2 \rightarrow CH_2Cl_2 + Cl\bullet$$

The likelihood of further substitution beyond the formation of CH_3Cl can be reduced if an excess of methane is used.

Similar free-radical substitution reactions can also occur with fluorine and bromine.

3.3.3 Halogenoalkanes

The halogenoalkanes are the homologous series of compounds with the general formula $C_nH_{2n+1}X$ where X is a halogen, i.e. F, Cl, Br or I, for example:

- CH_3CH_2Cl chloroethane

- $CH_3CHBrCH_3$ 2-bromopropane.

3.3.3.1 Nucleophilic substitution

Halogen atoms are **electronegative** (see Table 6) so that carbon–halogen bonds in halogenoalkanes are polar. The electrons in the C—X bond are attracted towards the halogen atom, which gains a slight negative charge, δ-, leaving the carbon atom electron-deficient or with a slight positive charge, δ+.

The δ+ carbon is then susceptible to attack by nucleophiles; i.e. negative ions or molecules with a lone pair of electrons. When nucleophilic attack occurs, the carbon–halogen bond breaks and a halide ion is released. The nucleophile replaces the halogen atom in a substitution reaction. The mechanism of this reaction (shown in Fig 29) is called **nucleophilic substitution**.

Table 6
Electronegativity values

Element	Electronegativity
C	2.5
F	4.0
Cl	3.0
Br	2.8
I	2.5

Fig 29
Nucleophilic substitution

$$\overset{..}{N}u \quad \overset{\delta+}{CH_3} \overset{\delta-}{Br} \longrightarrow \overset{+}{N}u-CH_3 + :Br^-$$

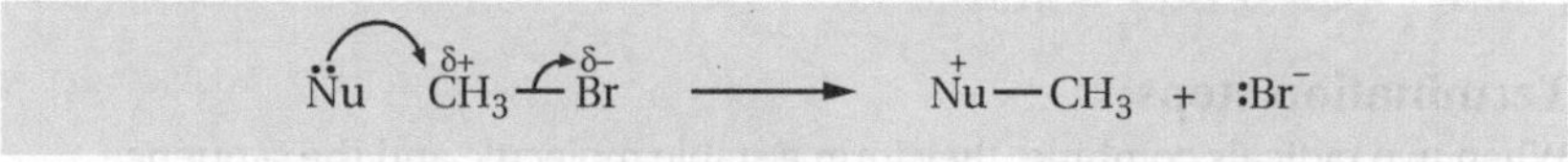

Essential Notes

$\overset{..}{N}u$ represents any nucleophile, an electron-pair donor.

The rate of such reactions is influenced by the strength of the carbon–halogen bond (see Table 7). Although the C—F bond is very polar, fluoroalkanes are very unreactive because the bond is so strong; chloroalkanes are also fairly slow to react. Carbon–bromine bonds, however, are more easily broken so that bromoalkanes react at a reasonable rate.

Bond	C—F	C—Cl	C—Br	C—I
Mean bond enthalpy/ kJ mol^{-1}	484	338	276	238

Table 7
Carbon–halogen bond strengths (mean bond enthalpies)

Nucleophilic substitution reactions

Nucleophilic substitution with hydroxide ions

When halogenoalkanes are warmed with aqueous sodium hydroxide or potassium hydroxide, alcohols are formed (see Fig 30).

Essential Notes

This process is sometimes called *hydrolysis.*

$$CH_3CH_2Br + OH^- \longrightarrow CH_3CH_2OH + Br^-$$
$$\text{ethanol}$$

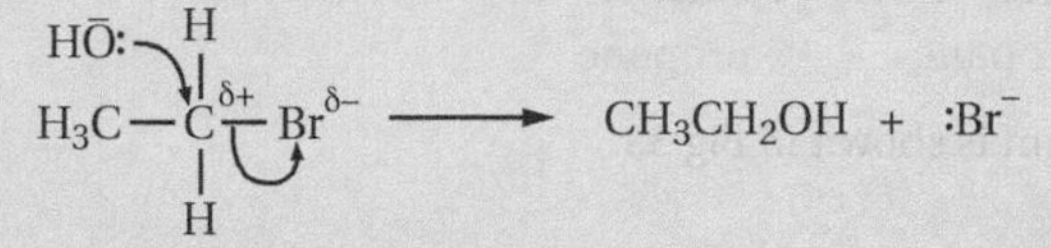

$$H\bar{O}: \begin{array}{c} H \\ | \\ H_3C-C^{\delta+}-Br^{\delta-} \\ | \\ H \end{array} \longrightarrow CH_3CH_2OH + :Br^-$$

Fig 30
Equation and mechanism for the formation of ethanol

Nucleophilic substitution with cyanide ions

When halogenoalkanes are warmed with an aqueous/alcoholic solution of potassium cyanide, nitriles are formed. For example, see Fig 31.

$$CH_3CH_2Br + CN^- \longrightarrow CH_3CH_2CN + Br^-$$
$$\text{propanenitrile}$$

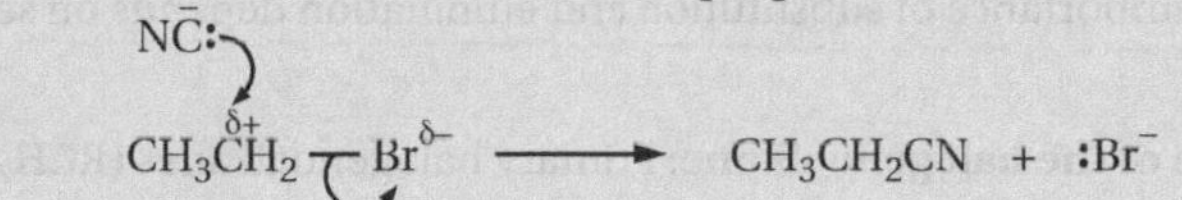

$$N\bar{C}: \\ CH_3\overset{\delta+}{C}H_2 - Br^{\delta-} \longrightarrow CH_3CH_2CN + :Br^-$$

Essential Notes

Nucleophilic substitution with cyanide ions adds an extra carbon to the chain. Compounds of the homologous series RCN are called nitriles.

Fig 31
Equation and mechanism for the formation of propanenitrile

Nucleophilic substitution with ammonia

When halogenoalkanes are warmed with an excess of ammonia in a sealed container, primary amines are formed. For example, bromoethane forms ethylamine:

$$CH_3CH_2Br + NH_3 \rightarrow CH_3CH_2NH_2 + HBr$$
$$\text{ethylamine}$$

Since the acid HBr immediately reacts with the base NH$_3$, this equation is more correctly written as:

$$CH_3CH_2Br + 2NH_3 \rightarrow CH_3CH_2NH_2 + NH_4Br$$

The mechanism has two steps, as shown in Fig 32.

Essential Notes

Reduction of a nitrile using hydrogen in the presence of a nickel catalyst also forms a primary amine (see *Collins Student Support Materials: A-Level year 2 – Organic and Relevant Physical Chemistry,* section 3.3.11.1).

Notes

Primary amines contain the NH$_2$ functional group.

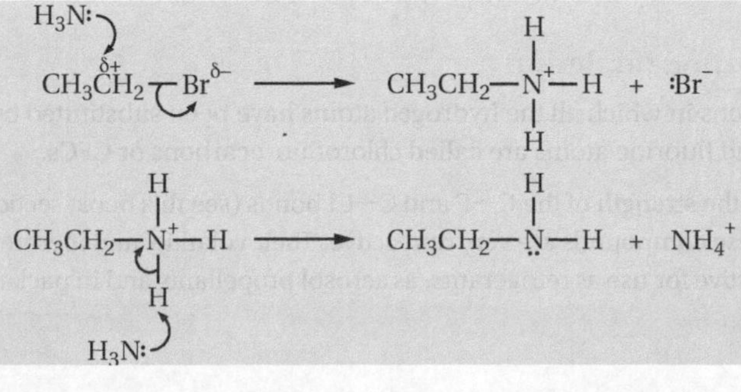

$$H_3N: \\ CH_3\overset{\delta+}{C}H_2 - Br^{\delta-} \longrightarrow CH_3CH_2-\overset{H}{\underset{H}{\overset{|}{N^+}}}-H + :Br^-$$

$$CH_3CH_2-\overset{H}{\underset{\underset{H_3N:}{|}}{\overset{|}{N^+}}}-H \longrightarrow CH_3CH_2-\overset{H}{\underset{}{\overset{|}{\ddot{N}}}}-H + NH_4^+$$

Fig 32
Mechanism for the formation of ethylamine

The excess of ammonia minimises the chance of further reaction of the primary amines to form secondary or tertiary amines, or quaternary ammonium salts (see *Collins Students Support Materials: A-Level year 2 – Organic and Relevant Physical Chemistry*, section 3.3.11.3).

3.3.3.2 Elimination

In the reaction between aqueous sodium hydroxide and a halogenoalkane, the hydroxide ion acts as a nucleophile, and an alcohol is formed by nucleophilic substitution. However, the hydroxide ion can also function as a base, so an alternative **elimination** reaction also takes place in which hydrogen and a halogen are eliminated from the halogenoalkane and an alkene is formed. For example:

$$CH_3CHBrCH_3 + OH^- \rightarrow CH_3CH{=}CH_2 + H_2O + Br^-$$
2-bromopropane   propene

The mechanism is shown in Fig 33.

Fig 33
Formation of propene by an elimination reaction

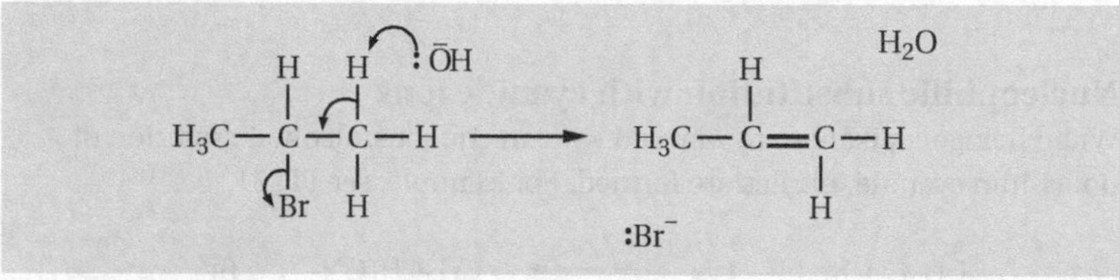

The relative importance of substitution and elimination depends on several factors:

- **Structure of the halogenoalkane**: Primary halogenoalkanes (RCH_2X) give predominantly **substitution** products, whereas tertiary halogenoalkanes (R_3CX) generally favour **elimination**. With secondary halogenoalkanes (R_2CHX), both **substitution** and **elimination** take place at the same time (concurrently).

- **Base strength of the nucleophile**: The likelihood of elimination increases as the base strength of the nucleophile increases.

- **Reaction conditions**: Higher reaction temperatures lead to a greater proportion of elimination.

In the reaction of 2-bromopropane with potassium hydroxide (see above), both substitution and elimination reactions occur together:

- elimination is favoured by hot ethanolic conditions
- substitution is favoured by warm aqueous conditions.

3.3.3.3 Ozone depletion

Hydrocarbons in which all the hydrogen atoms have been substituted by both chlorine and fluorine atoms are called chlorofluorocarbons or CFCs.

Because of the strength of the C—F and C—Cl bonds (see this book, section 3.3.3.1), these compounds are very unreactive. Their volatility and inertness made them attractive for use as refrigerants, as aerosol propellants and in packaging

Notes

Depending on the structure of the halogenoalkane, more than one alkene may be formed. Thus, for example, elimination from 2-bromobutane produces both but-1-ene and but-2-ene. The latter compound can, of course, exist as two E–Z stereoisomers; E-but-2-ene and Z-but-2-ene (see this book, section 3.3.1.3).

Essential Notes

Aqueous NaOH contains the nucleophile HO^- and NaOH in ethanol also contains the stronger base $CH_3CH_2O^-$.

materials such as expanded polystyrene. However, although this unreactive nature was beneficial in these uses, it was discovered that CFCs were a threat to the ozone layer in the upper atmosphere, and so they are no longer used.

Ozone, or trioxygen, O_3, is an **allotrope** of oxygen. It is formed when ultraviolet radiation from the sun breaks down oxygen molecules, O_2, into two oxygen atoms (radicals). These oxygen radicals then react with more oxygen to form ozone.

Ultraviolet radiation also decomposes ozone into an oxygen molecule and an oxygen radical, which can react with more oxygen to reform O_2 molecules.

$$O_2 \xrightarrow{\text{UV}} 2O\bullet$$
$$O\bullet + O_2 \rightarrow \bullet O_3$$
$$\bullet O_3 \xrightarrow{\text{UV}} O_2 + O\bullet$$
$$O\bullet + \bullet O_3 \rightarrow 2O_2$$

Strictly, oxygen should be shown as $\bullet O\bullet$ because it has two separate unpaired electrons.

The overall result of these reactions is that the presence of ozone in the upper atmosphere reduces the amount of harmful ultraviolet radiation from the sun that can reach the Earth's surface.

However, in the 1970s, it was discovered that the ozone 'layer' over Antarctica was considerably thinner than expected and the reason for this was found to be the reaction of chlorine radicals with ozone. These chlorine radicals are formed in the upper atmosphere when ultraviolet radiation causes the C—Cl bond in CFCs to break, for example:

$$CF_2Cl_2 \rightarrow Cl\bullet + \bullet CF_2Cl$$

The following is a possible sequence of reactions that then occur:

$$Cl\bullet + O_3 \rightarrow ClO\bullet + O_2$$
$$ClO\bullet + O_3 \rightarrow Cl\bullet + 2O_2$$

Chlorine radicals are removed in the first reaction but are re-formed in the second; overall, they are not used up and so are acting as a catalyst for the decomposition of ozone. This situation means that even small amounts of chlorine radicals can continue to deplete the ozone layer for many years.

As a result of this evidence, the use of CFCs was banned by international agreement (the Montreal protocol) and chemists have developed suitable replacements which do not contain chlorine. The most common of these are HFCs, which contain hydrogen, fluorine and carbon, such as CH_2FCF_3 (known as HFC 134a) and CH_3CHF_2 (known as HFC 152a) (see Fig 34). Since C—F bonds are stronger than C—Cl they are less likely to be broken by ultraviolet radiation. However, chemists will continue to monitor the atmosphere to check that these replacements do not themselves cause any long-term problems.

Fig 34
Examples of HFCs

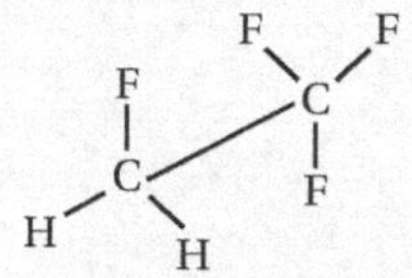

HFC 134a

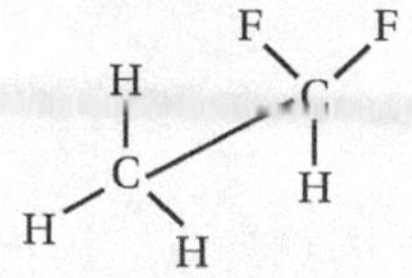

HFC 152a

3.3.4 Alkenes

Notes

The general formula for alkenes, C_nH_{2n}, can also represent cyclic alkanes. For example, C_6H_{12} can represent several isomeric hexenes and also cycloalkanes such as cyclohexane, methylcyclopentane and dimethylcyclobutane.

3.3.4.1 Structure, bonding and reactivity

The alkenes are a homologous series of hydrocarbons with the general formula C_nH_{2n}. The first three members are:

- ethene C_2H_4
- propene C_3H_6
- butene C_4H_8

Alkenes contain two hydrogen atoms fewer than their parent alkanes and are said to be **unsaturated**. This term is used because alkenes contain a double covalent bond between two carbon atoms (see Fig 35) and so they can become saturated by the addition of hydrogen.

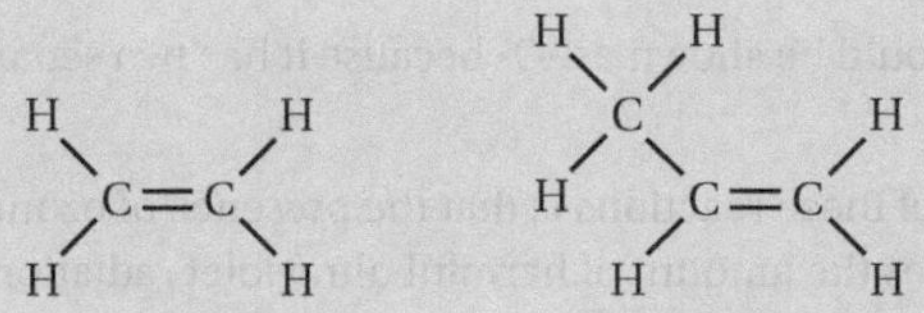

Fig 35
Structures of ethene and propene

ethene   propene

E–Z stereoisomerism

Carbon–carbon double bonds in alkenes cannot easily rotate because of the electron clouds present above and below the plane of the bond. When an alkene has two different groups at each end of the double bond, two different E–Z stereoisomers result. These compounds have the same structural formula, but the bonds within the molecule are arranged differently in space.

There are two types of stereoisomerism, E–Z **stereoisomerism**, discussed earlier in this book (section 3.3.1.3), and optical isomerism, which is discussed in *Collins Student Support Materials: A-Level year 2 – Organic and Relevant Physical Chemistry*, section 3.3.7.

3.3.4.2 Addition reactions of alkenes

The double bond in alkenes is an area of high electron density and is the cause of their reactivity. Alkenes can become saturated by the addition of small molecules across the double bond. In these reactions the C=C double bond becomes a C—C single bond.

The reaction of alkenes with small molecules such as HBr, H_2SO_4 or Br_2 occurs by an **electrophilic addition** mechanism:

- *electrophilic* because the electron-rich double bond attracts positive ions or electron-deficient atoms, and such species are called *electrophiles*
- *addition* because the electrophile joins on to the alkene to form one new molecule.

Notes

Electrophiles are positive ions or electron-deficient atoms and act as electron-pair acceptors; they seek electron-rich sites.

Hydrogen bromide

Alkenes react with hydrogen bromide in the gas phase, or in concentrated aqueous solution, to form bromoalkanes. For example ethene reacts to form bromoethane (see Fig 36).

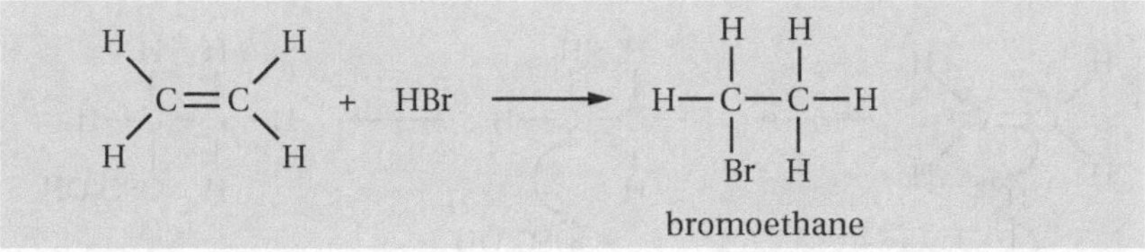

Fig 36
Equation for the reaction of ethene with HBr

The overall process involves electrophilic addition across a C=C double bond. The mechanism in the gas phase is shown in Fig 38.

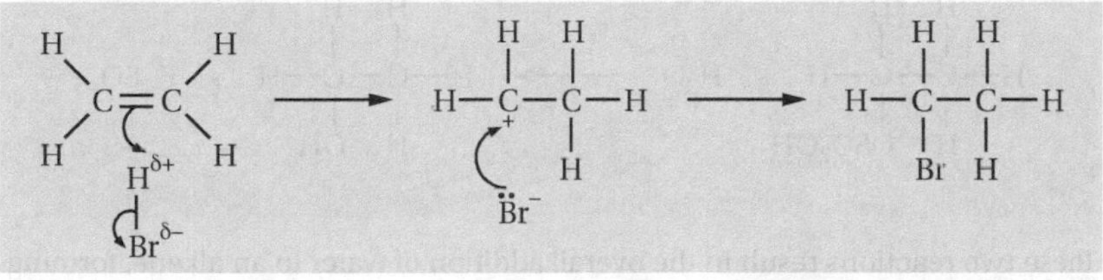

Fig 37
Elecrophilic addition mechanism for the reaction of ethene with HBr in the gas phase

The curly arrows in Fig 37 each represent the movement of a *pair* of electrons. Arrows start either at a lone pair or at the middle of a bond. Arrows end either between the atoms, where the new covalent bond forms, or as a lone pair on an atom.

- Bromine is more electronegative than hydrogen, so the HBr molecule is polar. The electron-deficient or $\delta+$ hydrogen atom acts as the electrophile (*electron-seeking species*).

- As electrons from the double bond move to form a new carbon-to-hydrogen bond with the $\delta+$ hydrogen, the electrons in the hydrogen–bromine bond shift towards the bromine atom and the H—Br bond breaks, releasing a bromide ion. The other carbon in the double bond becomes an electron-deficient **carbocation**.

- The bromide ion then acts as a nucleophile; it uses a lone pair of electrons to form a new bond with the positive carbon in the carbocation.

When the reaction is performed in aqueous solution, $H^+(aq)$ ions act as electrophiles in the first step of the mechanism. In the second step, bromide ions attack the carbocation, as in the mechanism shown in Fig 37.

Sulfuric acid

Alkenes are absorbed by cold, concentrated sulfuric acid to form alkyl hydrogensulfates. For example, Fig 38 shows how ethene forms ethyl hydrogensulfate.

Essential Notes

A **carbocation** is a species which contains a carbon atom that has a positive charge.

Fig 38
Equation and mechanism for the reaction of ethene with concentrated sulfuric acid

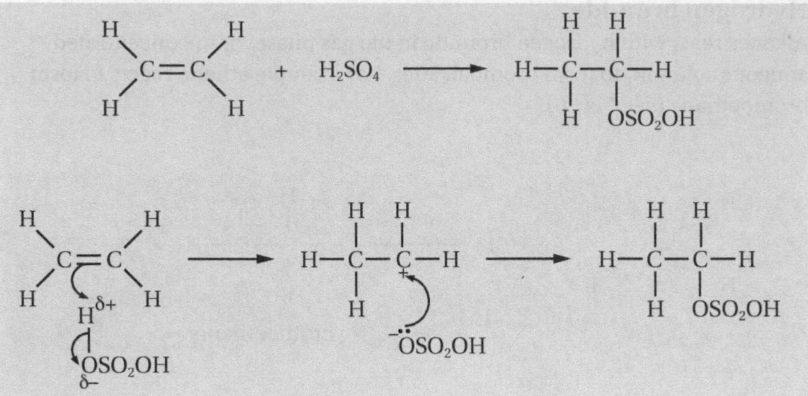

Notes

Ethanol is produced industrially by the direct hydration of ethene or by fermentation (see this book, section 3.3.5.1).

Warming ethyl hydrogensulfate in dilute sulfuric acid causes hydrolysis and produces ethanol by the reaction shown in Fig 40.

Fig 39
Hydrolysis of ethyl hydrogensulfate

Essential Notes

The amount of unsaturation in a margarine can be found by dissolving a sample in hexane and adding drops of bromine (*toxic and corrosive*). The more drops of bromine that are decolourised, the more unsaturation is present.

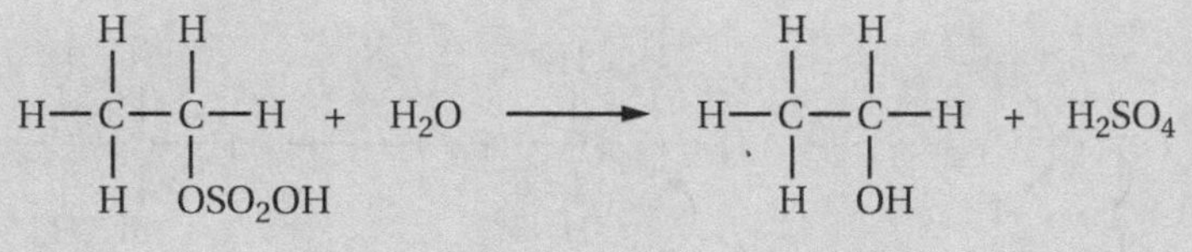

These two reactions result in the overall addition of water to an alkene, forming an alcohol.

Bromine a test for unsaturation

Alkenes decolourise solutions of bromine in water or in an organic solvent. Removal of the red–brown colour of bromine shows the presence of unsaturation, typically a C=C double bond. Alkanes do not react with bromine under these conditions, so that this reaction can be used to distinguish between alkanes and alkenes.

Ethene reacts with bromine to form a colourless, saturated product (Fig 40).

Fig 40
Equation for the reaction of ethene with bromine

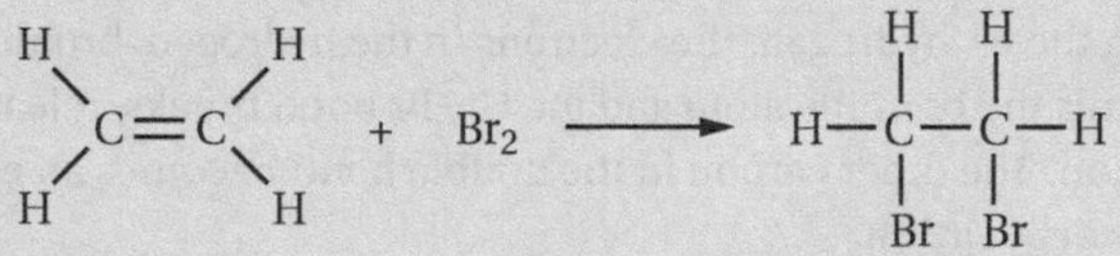

1,2-dibromoethane

Notes

In aqueous bromine (bromine water), an alternative product, see below, is also formed by attack of the nucleophile water on the carbocation.

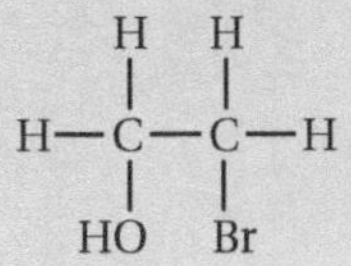

When molecules collide, the electron-rich region of the double bond repels the electrons in the bromine molecule and so induces a dipole in the bromine molecule. The electron-deficient or δ+ bromine atom is the electrophile in the mechanism (see Fig 41).

Fig 41
Electrophilic addition mechanism of bromine to ethene

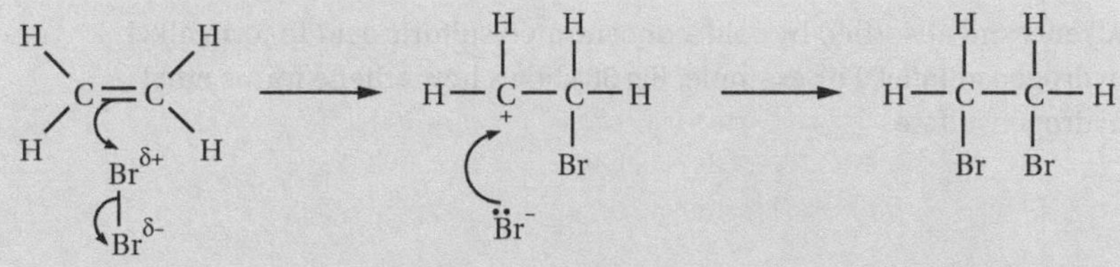

Electrophilic addition reactions to unsymmetrical alkenes

If the alkene is unsymmetrical, such as propene (see Fig 42), and the molecule being added is also unsymmetrical, such as hydrogen bromide (H—Br) or sulfuric acid (H—OSO$_2$OH), two possible products can form. The major product is the one formed via the *more stable* carbocation.

The order of stability of carbocations is tertiary (3°) > secondary (2°) > primary (1°), owing, to the inductive (electron-releasing) effect of the attached alkyl groups. The more alkyl groups around the carbocation, the more stable it is and the more likely it is to be formed.

The carbocations of C$_4$H$_9$$^+$ can have four possible structures (see Fig 43).

Fig 42
Structure of propene

Notes

A primary carbocation has one alkyl group attached to C$^+$.

A secondary carbocation has two alkyl groups attached to C$^+$.

A tertiary carbocation has three alkyl groups attached to C$^+$.

$$CH_3CH_2CH_2\overset{+}{C}H_2$$

$$(CH_3)_2CH\overset{+}{C}H_2$$

$$< \quad CH_3CH_2\overset{+}{C}HCH_3 \quad < \quad (CH_3)_3\overset{+}{C}$$

primary   secondary   tertiary

⟶ increasing stability of carbocations ⟶

Fig 43
Possible structures of C$_4$H$_9$$^+$

Propene reacts with hydrogen bromide to form mostly a secondary carbocation. This species then reacts with the bromide ion to produce 2-bromopropane (see Fig 44).

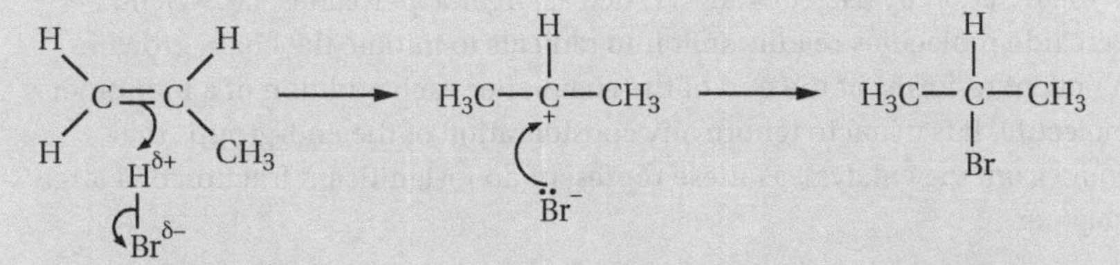

Fig 44
Mechanism of the reaction of propene with hydrogen bromide to form 2-bromopropane, the major product

Overall equation: $H_2C{=}CHCH_3 + HBr \longrightarrow CH_3CHBrCH_3$

2-bromopropane

A little 1-bromopropane will also be formed as the minor product via the less stable primary carbocation, by the mechanism shown in Fig 45.

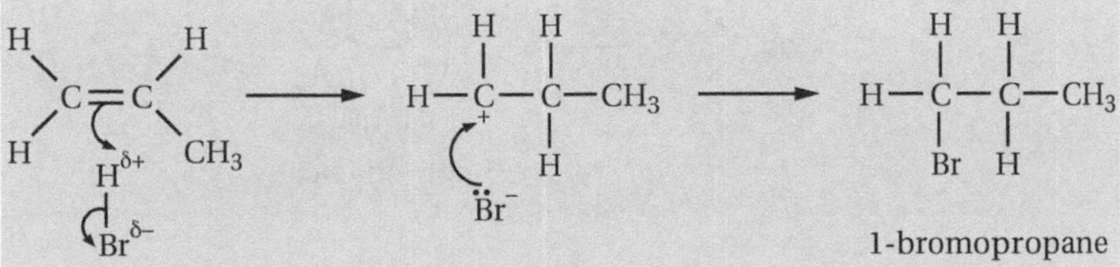

Fig 45
Formation of 1-bromopropane, the minor product

1-bromopropane

The reaction of propene with concentrated sulfuric acid produces mainly 2-propyl hydrogensulfate, via the secondary carbocation. Hydrolysis of this hydrogensulfate forms propan-2-ol (see Fig 46).

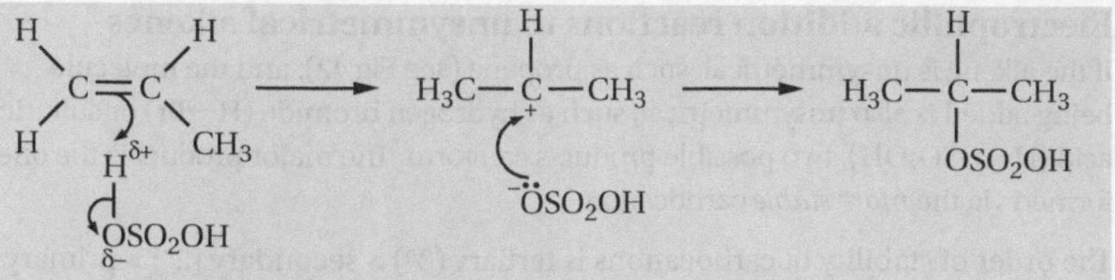

Fig 46
Reaction of propene with concentrated sulfuric acid

$$CH_3CH(OSO_2OH)CH_3 + H_2O \longrightarrow CH_3CH(OH)CH_3 + H_2SO_4$$
propan-2-ol

Notes

The bond energies of C=C and C—C bonds are 612 and 348 kJ mol^{-1}, respectively. Polymerisation, which involves forming two C—C bonds from one C=C bond, is therefore an exothermic process.

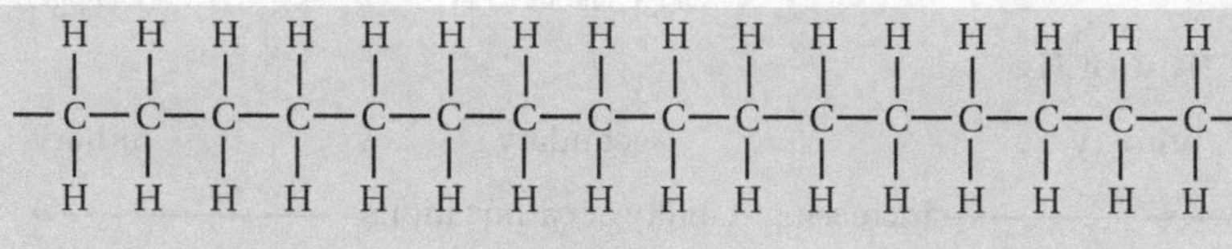

3.3.4.3 Addition polymers

Alkene molecules link together in the presence of a catalyst to form addition polymers which are saturated, such as poly(ethene). A section of the polymer (formed from eight ethene molecules) is shown in Fig 47.

Fig 47
Structure of poly(ethene)

Addition polymers are also known as chain-growth polymers. They are formed by the addition of monomers to the end of a growing chain. The end of the chain is reactive because it is a radical which is formed at the beginning of the reaction by use of catalysts such as organic peroxides, i.e. ROOR. Peroxide molecules readily split into radicals to initiate the chain growth. A radical re-forms at the end of the chain after each addition of a monomer molecule. It is usual to ignore any consideration of the end-groups that come from the catalyst, as these represent an insignificant fraction of a large polymer.

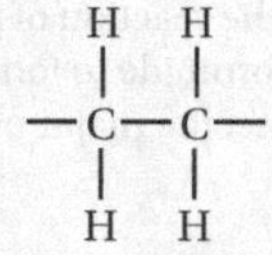

Fig 48
The repeating unit of poly(ethene)

Polymers formed from alkenes are usually represented using a repeating unit, such as that for ethene shown in Fig 48.

The polymerisation of ethene can therefore be represented by the equation in Fig 49.

Notes

n represents a large whole number, which is the number of individual molecules (monomers) joining together to form the polymer.

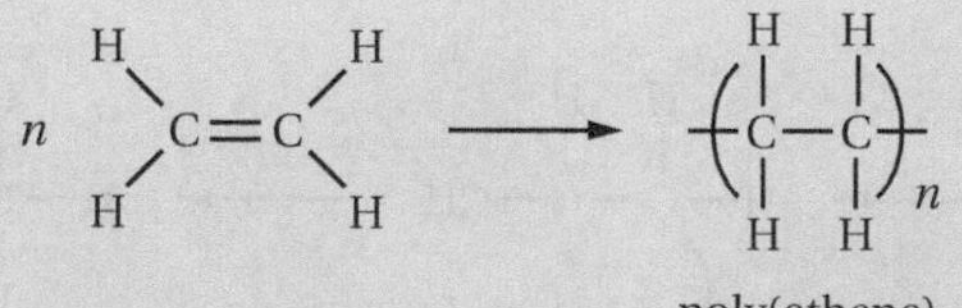

poly(ethene)

Fig 49
The formation of poly(ethene)

Polymers can be formed from monomers, in which some or all of the hydrogen atoms in ethene have been replaced. One example is poly(propene) (see Fig 50).

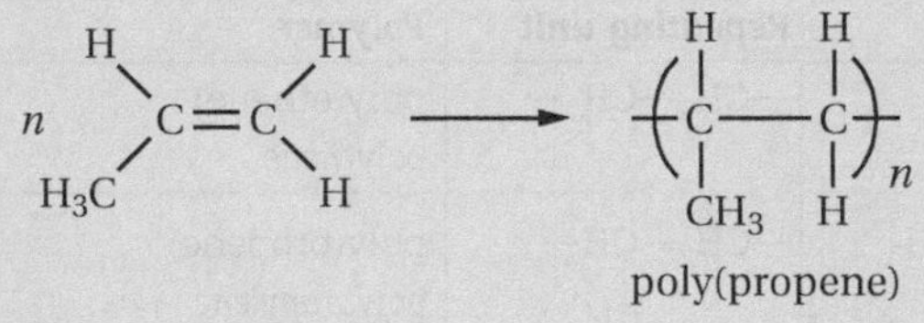

Fig 50
Formation of poly(propene)

IUPAC rules for the naming of addition polymers use the monomer name in brackets, preceded by 'poly'. For example, ethene polymerisation produces the polymer poly(ethene), and propene polymerisation produces poly(propene). Using this system, the polymer from chloroethene is called poly(chloroethene), but its non-IUPAC name, polyvinyl chloride (PVC) is still widely used.

Unlike alkenes, polyalkenes are saturated and therefore are unreactive, like simple alkanes, owing to strong covalent bonds between atoms and a lack of bond polarity. Polyalkenes are also non-biodegradable. However, polymers of this kind are, nevertheless, highly flammable. Methods of disposal of polymers are considered in *Collins Student Support Materials*: *A-Level year 2 – Organic and Relevant Physical Chemistry*, section 3.3.12.2.

There is no cross-linking between individual polyalkene chains so, as they are non-polar, there are only weak intermolecular van der Waals' forces of attraction between the chains. Long polymer chains have very large numbers of van der Waals' forces of attraction between them that strongly hold the chains together.

Poly(ethene) is used as packaging, especially as thin film and as 'plastic' bags.

Poly(propene) is very versatile. It is used to make rigid containers and objects such as car bumpers. It can also be made into fibres which are used as the backing for carpets and in thermal clothing.

Poly(chloroethene)/PVC has a range of uses; examples are window frames, guttering, plumbing and leather-look fabrics. This compound often has a plasticiser added to make it softer and more flexible, but there have been environmental concerns about some of these additives. For this reason, domestic cling film is increasingly made from poly(ethene) rather than PVC with added plasticiser.

Some polymers, such as poly(propene), can be recycled. Containers made of poly(propene) are collected, cleaned and cut into small pieces. The plastic is then melted and remoulded into a new object or extruded and spun into fibres.

Over time, our knowledge and understanding of the production and properties of polymers have developed, leading to an increased ability to control closely the length of polymer chains. Our ability to tailor the properties of polymers for specific uses and applications in industry has increased with the use of co-polymers (made by using two different monomers).

Some illustrative examples of addition polymers are given in Table 8.

Table 8
Some addition polymers and their
typical uses

Monomer	Repeating unit	Polymer	Typical uses
$CH_2\!=\!CH_2$	$-CH_2-CH_2-$	poly(ethene) *polythene*	film, bags
$CH_2\!=\!CHCH_3$	$-CH_2-CH-$ $\qquad\;\; CH_3$	poly(propene) *polypropylene*	moulded plastics, fibres
$CH_2\!=\!CHC_6H_5$	$-CH_2-CH-$ $\qquad\;\; C_6H_5$	poly(phenylethene) *polystyrene*	packaging, insulation
$CH_2\!=\!CHCl$	$-CH_2-CH-$ $\qquad\;\; Cl$	poly(chloroethene) *poly(vinyl chloride), (PVC)*	pipes, flooring
$CH_2\!=\!CHCN$	$-CH_2-CH-$ $\qquad\;\; CN$	poly(propenenitrile) *poly(acrylonitrile)*	fibres
$CH_2\!=\!CHOCOCH_3$	$-CH_2-CH-$ $\qquad\;\; OCOCH_3$	poly(ethenyl ethanoate) *poly(vinyl acetate)*	paints, adhesives
$CH_2\!=\!C-COOCH_3$ $\quad\;\; CH_3$	$\qquad\;\; COOCH_3$ $-CH_2-C-$ $\qquad\;\; CH_3$	poly(methyl 2-methylpropenoate) *poly(methyl methacrylate)*	glass replacement (Perspex), baths
$CH_2\!=\!C-COOCH_3$ $\quad\;\; CN$	$\qquad\;\; COOCH_3$ $-CH_2-C-$ $\qquad\;\; CN$	poly(methyl 2-cyanopropenoate) *poly(methyl cyanoacrylate)*	super-glue
$CF_2\!=\!CF_2$	$-CF_2-CF_2-$	poly(tetrafluoroethene) *poly(tetrafluoroethylene), (PTFE)*	non-stick surfaces, non-lubricated bearings

3.3.5 Alcohols

The alcohols are the homologous series with the general formula $C_nH_{2n+1}OH$. They all contain the functional group $-OH$, which is called the hydroxyl group. The first two alcohols are:

* methanol $\qquad\qquad\qquad$ CH_3OH
* ethanol $\qquad\qquad\qquad\;\;$ CH_3CH_2OH.

This section also includes three other homologous series of organic compounds: aldehydes, ketones and carboxylic acids; these compounds all contain the carbonyl group $C\!=\!O$. The naming of these compounds is illustrated in Table 9 and follows the same rules as were explained in section 3.3.1.1.

Table 9
Alcohols, carbonyl compounds and
carboxylic acids

Homologous series	Name: prefix or suffix	Functional group	Example
alcohols	suffix -ol prefix hydroxy-	—OH	ethanol CH_3CH_2OH 2-hydroxy-propanoic acid $CH_3CH(OH)COOH$
aldehydes	suffix -al	$-C\overset{O}{\underset{H}{<}}$	ethanal CH_3CHO
ketones	suffix -one prefix oxo-	$>C=O$	propanone CH_3COCH_3 3-oxobutanoic acid CH_3COCH_2COOH
carboxylic acids	suffix -oic acid	$-C\overset{O}{\underset{OH}{<}}$	ethanoic acid CH_3COOH

3.3.5.1 Alcohol production

Direct hydration
Ethanol is produced industrially by the direct hydration of ethene using steam and a phosphoric acid catalyst at 300 °C and 6.5×10^3 kPa pressure:

$$C_2H_4(g) + H_2O(g) \rightleftharpoons C_2H_5OH(g)$$

Direct hydration is currently preferred for the production of ethanol for industrial use in the UK. However, as this method uses ethene as a raw material, it may become less popular compared to fermentation when oil supplies begin to run out. Table 10 compares the two methods of production.

Fermentation
Alcohol is also produced by the process of fermentation, which uses living yeast cells to convert sugars such as glucose into ethanol and carbon dioxide:

$$C_6H_{12}O_6 \xrightarrow{\text{yeast}} 2C_2H_5OH + 2CO_2$$

At low temperatures the reaction is slow, as the enzymes (natural catalysts) in yeast are inactivated; at high temperatures, the yeast cannot survive. The process is therefore normally carried out at a compromise temperature of about 35 °C.

Method	Rate of reaction	Quality of product	Raw material	Type of process
hydration	fast	pure	ethene from oil (a finite resource)	continuous (cheap on manpower) (expensive equipment)
fermentation	slow	impure	sugars (a renewable resource)	batch (expensive on manpower) (cheap equipment)

Essential Notes

The most common alcohol is ethanol, which is present in 'alcoholic' drinks.

Essential Notes

Fermentation produces an aqueous solution of ethanol at a concentration of between 3% and 15%. Beers usually contain about 3–7% ethanol and wines about 9–15%. Fermentation rarely produces higher concentrations of ethanol because high alcohol concentrations kill the yeast. More concentrated solutions in spirits such as whisky, brandy or gin, which are about 40% ethanol, are made by distillation of the fermented products.

Table 10
Comparison of methods used to produce ethanol

Ethanol produced by fermentation is called a **biofuel**. Such a fuel is produced from plants or biomass (material derived from plants), which is renewable because the plant can be grown again quickly. These fuels contrast with fossil fuels, which took millions of years to form and are not renewable.

Combustion of any carbon-containing fuel will eventually increase the amount of carbon dioxide in the atmosphere. However, ethanol as a biofuel can be considered to be **carbon neutral**. This is because the amount of carbon dioxide produced in its combustion, added to the carbon dioxide released during its formation by fermentation, equals the amount of carbon dioxide removed from the atmosphere during photosynthesis, which produces the sugars used in the fermentation that produces the ethanol (see Table 11).

Table 11
Carbon-neutral status of ethanol produced by fermentation

Removal of carbon dioxide from the atmosphere	Release of carbon dioxide into the atmosphere
Photosynthesis	Fermentation
$6CO_2 + 6H_2O \rightarrow C_6H_{12}O_6 + 6O_2$	$C_6H_{12}O_6 \rightarrow 2C_2H_5OH + 2CO_2$
6 mol of CO_2 removed per 1 mol of sugar formed	2 mol of CO_2 released per 1 mol of sugar fermented
	Combustion
	$2C_2H_5OH + 6O_2 \rightarrow 4CO_2 + 6H_2O$ 4 mol of CO_2 released during the combustion of the 2 mol of ethanol formed per 1 mol of sugar fermented
Total: 6 moles of CO$_2$ removed	**Total: 6 moles of CO$_2$ released**

Notes

Although Table 11 seems to indicate that ethanol can be considered to be carbon neutral, the production and processing steps require energy. The production of this energy may involve the release of additional carbon dioxide.

Although, as shown in Table 11, bioethanol is in theory carbon neutral, this approach does not take into account the carbon dioxide emissions associated with growing, harvesting and transporting the crops, or producing the ethanol. Therefore, overall, more carbon dioxide is emitted than is absorbed – which means that it contributes to global warming.

Also, there are concerns about the increased production of bioethanol and similar fuels. The land area used to grow crops suitable for fermentation may increase and can lead to further deforestation (as in the Amazon basin). Trees act as a natural 'store' of carbon dioxide, but this carbon dioxide is released after deforestation when timber is burned.

Also, the use of land to grow crops to produce these fuels in developing countries has already reduced the land available for growing crops for much-needed food.

3.3.5.2 Oxidation of alcohols

Alcohols can be classified as primary, secondary or tertiary, depending on the carbon skeleton to which the hydroxyl group is attached, as shown in Fig 51, where R is any alkyl group.

Many reactions of the OH functional group are the same in all alcohols, independent of where the OH group is attached to the carbon skeleton. However, the three types of alcohol differ in their reactions with oxidising agents such as acidified potassium dichromate(VI).

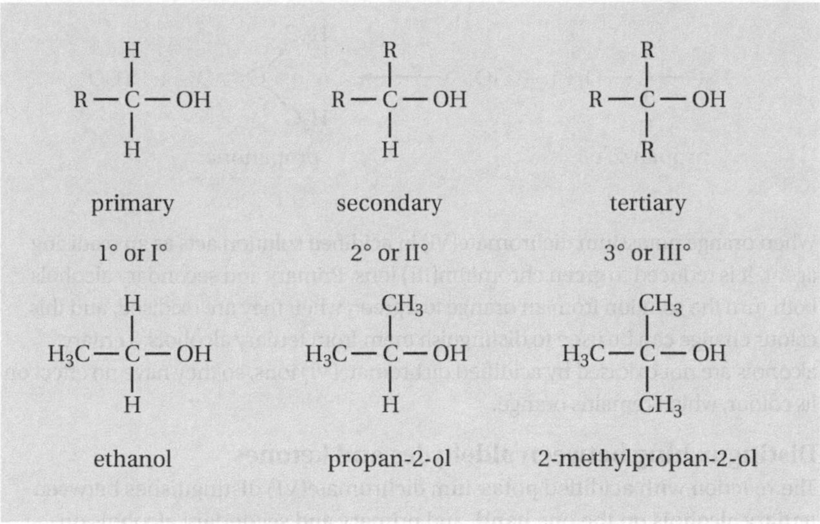

Fig 51
Classification of alcohols
(R represents any alkyl group)

primary   secondary   tertiary

1° or I°   2° or II°   3° or III°

ethanol   propan-2-ol   2-methylpropan-2-ol

Primary alcohols are oxidised first to aldehydes, such as ethanal, as shown in Fig 52.

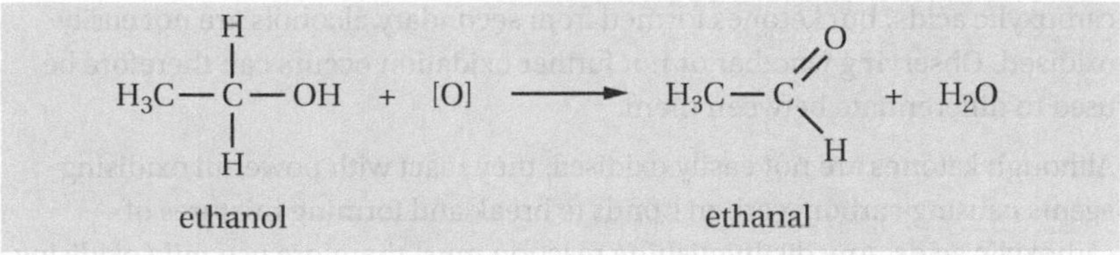

$$H_3C-\overset{\displaystyle H}{\underset{\displaystyle H}{C}}-OH \;+\; [O] \;\longrightarrow\; H_3C-\overset{\displaystyle O}{C}{\diagdown}_H \;+\; H_2O$$

ethanol   ethanal

Fig 52
Oxidation of ethanol

> **Notes**
>
> The use of [O] to represent the oxidant is an allowed simplification in this and other equations showing the oxidation of organic compounds. The equations, however, must still balance.

An aldehyde still has one hydrogen atom attached to the carbonyl carbon, and so it can be oxidised one step further to a carboxylic acid (see Fig 53).

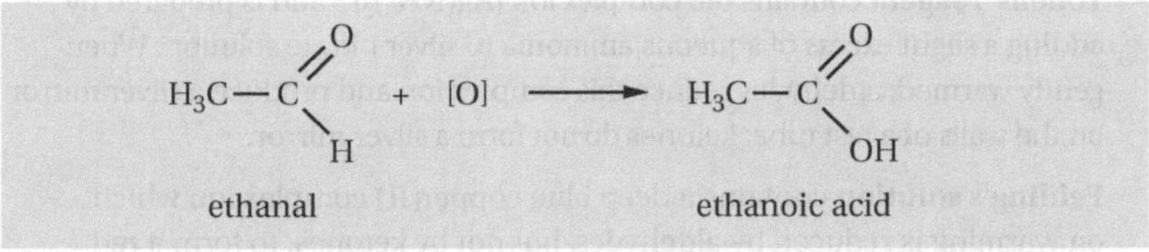

$$H_3C-\overset{\displaystyle O}{C}{\diagdown}_H \;+\; [O] \;\longrightarrow\; H_3C-\overset{\displaystyle O}{C}{\diagdown}_{OH}$$

ethanal   ethanoic acid

Fig 53
Oxidation of ethanal

In practice, a primary alcohol such as ethanol is dripped into a warm solution of acidified potassium dichromate(VI). The aldehyde, ethanal, is formed and immediately distils off, thereby preventing further oxidation to ethanoic acid, because the boiling point of ethanal (23 °C) is much lower than that of either the original alcohol ethanol (78 °C) or of ethanoic acid (118 °C). Both the alcohol and the acid have higher boiling points because of **hydrogen bonding**.

If oxidation of ethanol to ethanoic acid is required, the reagents must be heated together under **reflux** to prevent escape of the aldehyde before it can be oxidised further.

Secondary alcohols are oxidised to ketones (see Fig 54). These have no hydrogen atoms attached to the carbonyl carbon and so cannot easily be oxidised further.

Essential Notes

Alcoholic drinks such as wine and beer that are left exposed to air go sour. This is because the ethanol present is oxidised to ethanoic acid by oxygen using enzymes from bacteria. However, the bacteria cannot tolerate ethanol concentrations greater than about 20%, so the ethanol in high alcohol drinks is not oxidised. This applies to all spirits and also to fortified wines such as sherry and port.

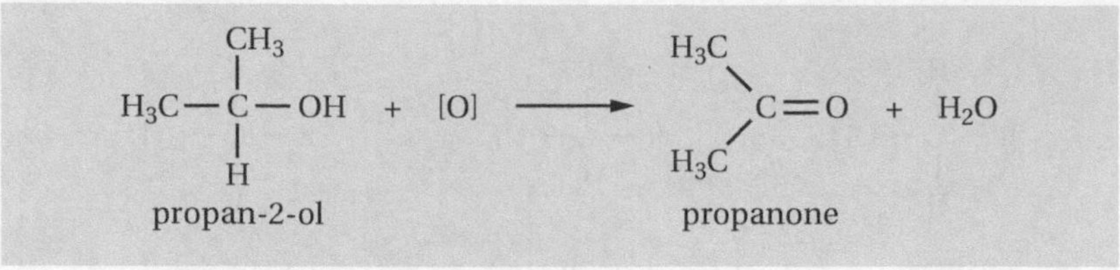

Fig 54
Oxidation of propan-2-ol

When orange potassium dichromate(VI) in acidified solution acts as an oxidising agent, it is reduced to green chromium(III) ions. Primary and secondary alcohols both turn the solution from an orange to a green when they are oxidised, and this colour change can be used to distinguish them from tertiary alcohols. Tertiary alcohols are not oxidised by acidified dichromate(VI) ions, so they have no effect on its colour, which remains orange.

Distinguishing between aldehydes and ketones

The reaction with acidified potassium dichromate(VI) distinguishes between tertiary alcohols on the one hand, and primary and secondary alcohols on the other, but cannot distinguish between primary and secondary alcohols. However, these can be differentiated by further tests on their oxidation products. Aldehydes formed from primary alcohols are easily oxidised to carboxylic acids, but ketones formed from secondary alcohols are not easily oxidised. Observing whether or not further oxidation occurs can therefore be used to differentiate between them.

Although ketones are not easily oxidised, they react with powerful oxidising agents causing carbon–carbon bonds to break and forming mixtures of carboxylic acids. Any distinguishing reaction must therefore use *mild* oxidising agents to prevent this bond-breaking from occurring. The following are two such examples:

- **Tollens' reagent** contains the complex ion $[Ag(NH_3)_2]^+$ and is prepared by adding a slight excess of aqueous ammonia to silver nitrate solution. When gently warmed, aldehydes reduce this complex ion and produce a silver mirror on the walls of a test tube; ketones do not form a silver mirror.

- **Fehling's solution** contains a deep blue copper(II) complex ion which on warming is reduced by aldehydes, but not by ketones, to form a red precipitate of copper(I) oxide, Cu_2O.

When aldehydes act as reducing agents in these reactions, they are oxidised to carboxylic acids as shown by the equation:

$$RCHO + [O] \rightarrow RCOOH$$

3.3.5.3 Elimination

Alcohols with a hydrogen atom on the carbon next to the OH group can be dehydrated to alkenes when heated to about 180 °C with concentrated sulfuric or concentrated phosphoric acid. The reaction is an acid-catalysed elimination.

Examples of such elimination reactions are shown in Figs 55–57.

Essential Notes

Tertiary alcohols are not easily oxidised.

Essential Notes

If a large excess of ammonia is used, this reacts with the aldehyde leaving none to react with the $[Ag(NH_3)_2]^+$ ion. Hence, a silver mirror will not be seen.

Notes

Some alcohols may form more than one alkene on dehydration, e.g. butan-2-ol can form but-1-ene and but-2-ene (see also elimination from halogenoalkanes in this book, section 3.3.3.2).

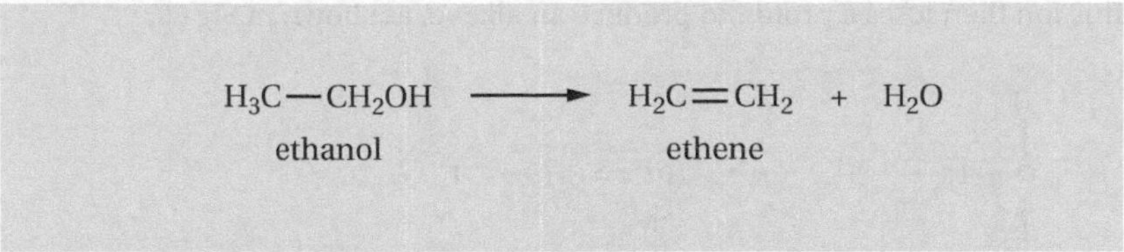

$$H_3C-CH_2OH \longrightarrow H_2C=CH_2 + H_2O$$
$$\text{ethanol} \qquad\qquad \text{ethene}$$

Fig 55
Dehydration of ethanol

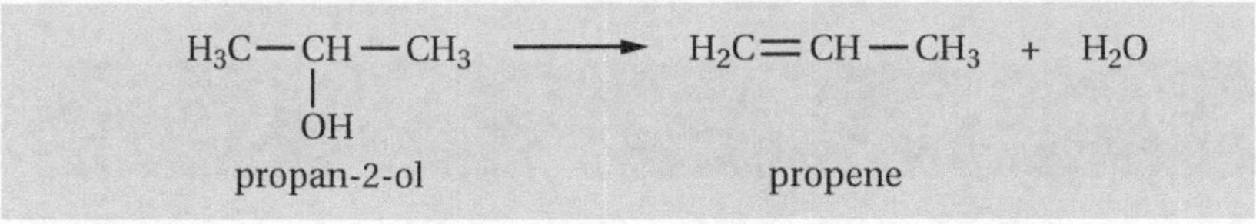

$$H_3C-CH-CH_3 \longrightarrow H_2C=CH-CH_3 + H_2O$$
$$\quad\;\; | $$
$$\quad\;\; OH$$
$$\text{propan-2-ol} \qquad\qquad \text{propene}$$

Fig 56
Dehydration of propan-2-ol

cyclohexanol   cyclohexene

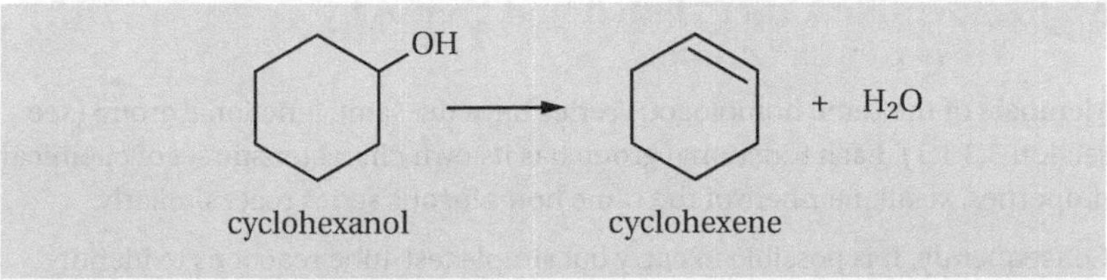

Fig 57
Dehydration of cyclohexanol

The alkenes produced, such as ethene and propene, can be used as the starting materials (monomers) in addition polymerisation reactions (see this book, section 3.3.4.3). Hence, this dehydration reaction provides an alternative route to addition polymers compared with the usual method starting from oil.

The two routes are:

$$\text{alcohols} \xrightarrow{\text{dehydration}} \text{alkenes} \longrightarrow \text{polymers}$$

$$\text{crude oil} \xrightarrow[\text{distillation}]{\text{fractional}} \text{alkanes} \xrightarrow{\text{cracking}} \text{alkenes} \longrightarrow \text{polymers}$$

The route from alcohols may use renewable starting materials, whereas production from crude oil uses up vital natural resources.

Elimination mechanism

The mechanism is an elimination. H^+ ions are used in the first step but released in the final step; i.e. the reaction is acid catalysed.

Consider the dehydration of propan-2-ol to propene.

In the first step, the alcohol is protonated, as shown in Fig 58.

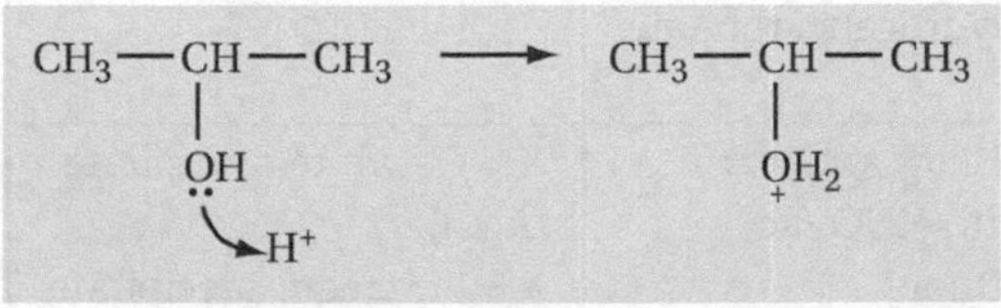

$$CH_3-CH-CH_3 \longrightarrow CH_3-CH-CH_3$$
$$\qquad | \qquad\qquad\qquad\qquad |$$
$$\qquad OH \qquad\qquad\qquad\qquad OH_2$$
$$\qquad\;\; H^+$$

Fig 58
Protonation

The protonated alcohol then loses water to form a carbocation, as shown in Fig 59.

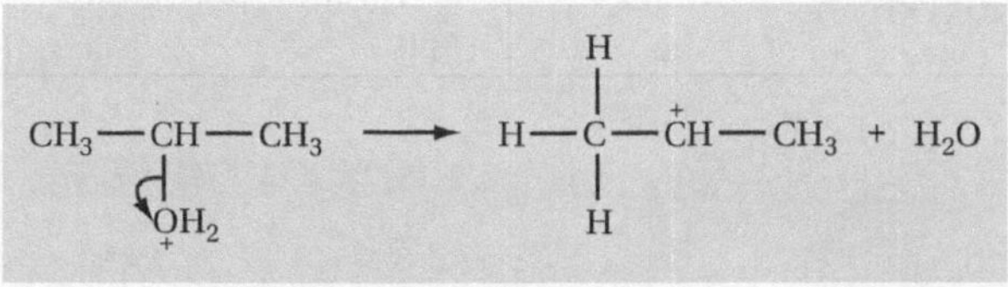

$$CH_3-CH-CH_3 \longrightarrow H-\overset{\overset{\displaystyle H}{|}}{\underset{\underset{\displaystyle H}{|}}{C}}-\overset{+}{C}H-CH_3 + H_2O$$
$$\qquad | $$
$$\qquad OH_2$$

Fig 59
Loss of water

> **Notes**
> The distillation of a product from a reaction is a required practical activity.

45

This ion then loses a proton to produce an alkene, as shown in Fig 60.

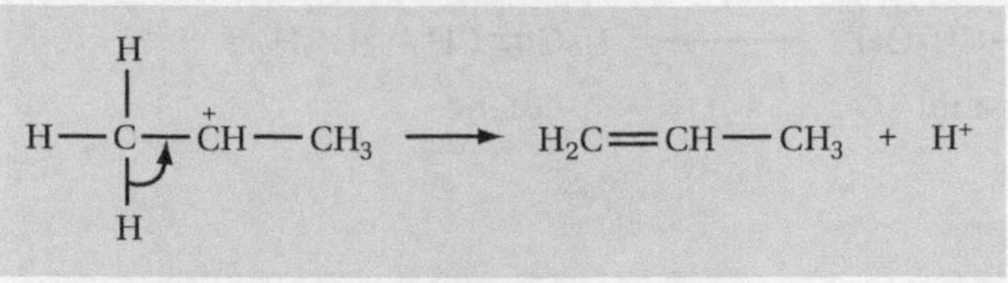

Fig 60
Loss of a proton

$$H-\overset{\overset{\displaystyle H}{|}}{\underset{\underset{\displaystyle H}{|}}{C}}-\overset{+}{C}H-CH_3 \longrightarrow H_2C=CH-CH_3 + H^+$$

3.3.6 Organic analysis

3.3.6.1 Identification of functional groups by test-tube reactions

Members of the same homologous series have the same functional group (see section 3.3.1.1). Each functional group has its own characteristic set of chemical properties, so all members of the same homologous series react similarly.

Consequently, it is possible to carry out simple test-tube reactions to identify functional groups.

Some specific tests for functional groups are set out in Table 12:

Table 12
Some reactions of functional groups

Functional group	Test	Observation
alkene $C=C$	shake with bromine water	alkenes decolourise bromine water
alcohol $R-OH$	warm with acidified potassium dichromate(VI), $K_2Cr_2O_7$	primary and secondary alcohol: orange solution to green solution tertiary alcohol: no change
aldehyde $R-CHO$	warm with Fehling's solution or warm with Tollens' reagent	blue solution to red precipitate silver mirror forms
carboxylic acid $R-COOH$	add sodium carbonate, Na_2CO_3 or sodium hydrogencarbonate, $NaHCO_3$	effervescence, or it fizzes (as CO_2 is given off)
halogenoalkane $R-X$	add NaOH(aq) and warm acidify with HNO_3(aq) add $AgNO_3$(aq)	X = Cl: white precipitate (AgCl) X = Br: cream precipitate (AgBr) X = I: yellow precipitate (AgI)

Notes
Testing for alcohol, aldehyde, alkene and carboxylic acid functional groups is a required practical activity.

3.3.6.2 Mass spectrometry

When a sample of an organic compound (M) is introduced into a mass spectrometer, the peak at the maximum value of m/z (see *Collins Student Support Materials: AS/A-Level year 1 – Inorganic and Relevant Physical Chemistry*, section 3.1.1.2) corresponds to the **molecular ion, $M^{+\bullet}$**. The value of m/z for this ion is equal to the **relative molecular mass** (M_r). ($m/z = M_r$ because $z = 1$ for a unipositive ion.)

Many organic molecules have the same **integral mass**. Thus, for example, to the nearest integer, the relative molecular masses of C_6H_{12}, C_5H_8O, $C_4H_4O_2$ and $C_4H_8N_2$ are identical ($M_r = 84$). However, if more precise **relative atomic masses** are used, slightly different M_r values are obtained for these molecules (see Table 13).

High-resolution mass spectrometers are capable of measuring the precise m/z of any molecular ion. By comparing this experimental value with that calculated for each species having the same integral mass, it is possible to assign a molecular formula to an unknown ion. Modern high-resolution instruments possess computer programs designed to match precise masses with molecular formulae. It needs to be appreciated, of course, that a particular molecular formula, although unambiguous, can represent many isomers. For example, C_3H_6O is the molecular formula of nine different structural isomers.

Mass spectra reveal the presence of **isotopes**. Of particular interest are molecules containing bromine or chlorine, where two isotopes are present in substantial quantities. The ratio of the relative isotopic abundances of ^{35}Cl (75.8%) and ^{37}Cl (24.2%) is close to 3:1. Thus, for example, the mass spectra of monochloroalkanes exhibit molecular ions two mass units apart in a 3:1 intensity ratio due to the presence of $R^{35}Cl$ and $R^{37}Cl$, respectively.

Similarly, the mass spectra of monobromoalkanes show molecular ion peaks two mass units apart in a near to 1:1 intensity ratio, because ^{79}Br (50.7%) in $R^{79}Br$ and ^{81}Br (49.3%) in $R^{81}Br$ have approximately the same abundance (Fig 61). These distinctive features are helpful in revealing the presence of chlorine or bromine.

Notes

Note that the species $M^{+\bullet}$ formed in a mass spectrometer is a cation and also a radical, because a covalent bond has lost one of its two electrons.

Compound	Precise mass
C_6H_{12}	84.1590
C_5H_8O	84.1161
$C_4H_4O_2$	84.0732
$C_4H_8N_2$	84.1194

Table 13
Precise masses of four compounds with $m/z = 84$

Essential Notes

Exercise: Try to work out the nine structural isomers of C_3H_6O, some of which are cyclic.

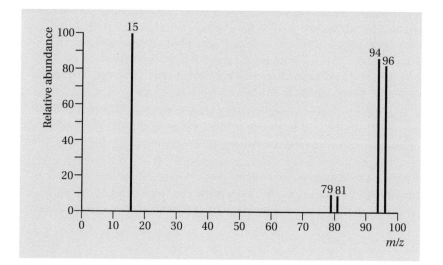

Fig 61
Simplified mass spectrum of bromomethane, CH_3Br

Notes

Frequency (f) is related to wavelength (λ) by the expression $f = c/\lambda$, where c is the speed of light; f is given in Hz (s^{-1}). By convention, however, infrared band positions are quoted as reciprocal wavelengths ($1/\lambda$), alternatively called wavenumbers, with units cm^{-1}. wavenumbers are directly related to energy ($E = hf$, where h is the Planck constant).

Essential Notes

The wavenumber range 400 to 4000 cm^{-1} corresponds to an absorption of energy between 4.8 and 48 kJ mol^{-1}. The energy involved is much less than that required to break covalent bonds.

Table 14
Regions in the infrared spectrum

Notes

Note that appropriate infrared data will be provided in examination questions and on the reverse side of the Periodic Table provided.

Notes

Absorptions due to various kinds of C—H bending in saturated groups appear characteristically at about 1460 cm^{-1} and 1370 cm^{-1}.

3.3.6.3 Infrared spectroscopy

The entire infrared spectrum of an organic compound provides a unique **molecular fingerprint**. Thus, if two samples have identical infrared spectra, the compounds must be identical. Also, comparison can be made between the full infrared spectrum of a pure material and the full spectrum of a sample, when the presence of any impurities will show up as extra peaks.

Of more general use is the *identification of functional groups* in organic compounds. Most functional groups give rise to characteristic infrared absorptions which change little from one compound to another.

Absorption of infrared energy by a molecule causes *bond stretching* and *bond bending*, giving rise to the appearance of peaks at particular **wavenumbers**. The position of an infrared peak depends on the bond strength and on the masses of the atoms which are joined by the bond. Since bending involves less energy than stretching, for the same bonds, bending absorptions occur at lower wavenumbers than stretching modes. Strong bonds and light atoms vibrate at relatively high wavenumbers. Conversely, weak bonds and heavy atoms absorb at lower wavenumbers.

It is convenient to divide the infrared spectrum into four regions for purposes of structure interpretation (see Table 14).

Region/cm^{-1}	Absorptions
4000–2500	C—H, O—H, N—H
2500–2000	C≡C, C≡N
2000–1500	C=C, C=O, C=N
1500–400	C—C, C—O, C—N, C—X

The region below 1500 cm^{-1} is the **fingerprint region**, which is often complex owing to a wide variety of single-bond vibrations. Nearly all organic molecules absorb strongly close to 3000 cm^{-1}, owing to C—H stretching vibrations associated with alkyl groups. The position of the C—H stretch is shifted by the presence of adjacent multiple bonds (see Table 15). The aldehyde C—H stretch is distinctive.

Carbonyl containing compounds show a very characteristic strong absorption in the range 1600–1800 cm^{-1}, the actual value of which depends on the nature of the adjacent groups. For example, simple aldehydes and ketones absorb at about 1720 cm^{-1}, whereas esters are closer to 1740 cm^{-1}. Owing to hydrogen bonding, hydroxyl groups in alcohols tend to have broad O—H absorption bands around 3300 cm^{-1}; carboxylic acid groups have very broad O—H bands in the region of 3000 cm^{-1}.

The examination of an infrared absorption spectrum is a useful way of identifying particular functional groups, especially when used in conjunction with mass spectrometry or nuclear magnetic resonance spectroscopy. It is important to appreciate that much structural information can be derived from an infrared spectrum by noticing which characteristic absorptions are *absent*.

Illustrative infrared absorption spectra are shown for propan-2-ol (Fig 62), propanal (Fig 63) and propanoic acid (Fig 64). Real samples are rarely perfect and often contain trace amounts of water. In such cases, weak absorptions in the O—H region can be observed (see Fig 63).

Table 15
Some characteristic infrared absorptions due to bond stretching in organic molecules

Bond	Types of compound	Range/cm⁻¹
C—H	alkanes	2850–2960
	alkenes	3010–3095
	alkynes	3250–3300
	arenes	3030–3080
	aldehydes	2710–2730
O—H	alcohols (H-bonded)	3230–3550
	carboxylic acids (H-bonded)	2500–3000
N—H	amines	3320–3560
C—C	alkanes	750–1100
C=C	alkenes	1620–1680
C≡C	alkynes	2100–2250
C=O	alcohols, ethers, carboxylic acids, esters	1000–1300
C=O	aldehydes, ketones, carboxylic acids, esters	1680–1750
C—N	amines	1180–1360
C≡N	nitriles	2210–2260
C—Cl	haloalkanes	600–800
C—Br	haloalkanes	500–600

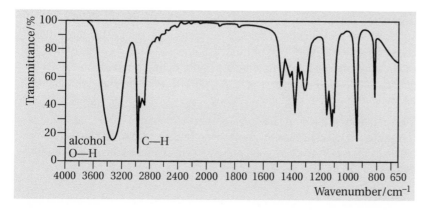

Fig 62
Infrared spectrum of propan-2-ol, $(CH_3)_2CHOH$

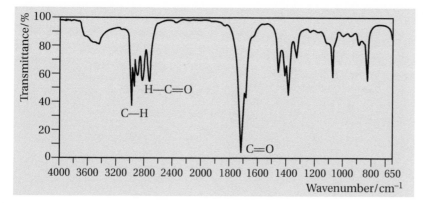

Fig 63
Infrared spectrum of propanal, CH_3CH_2CHO

Fig 64
Infrared spectrum of propanoic acid,
CH_3CH_2COOH

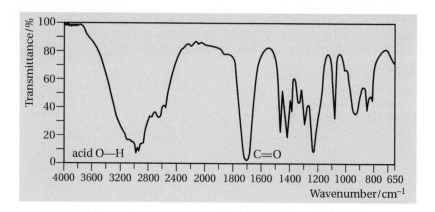

Absorption of infrared radiation and global warming

The gases carbon dioxide, methane and water vapour are often referred to as **greenhouse gases**. These gases and others in the atmosphere absorb infrared radiation emitted by the Earth that might otherwise escape into space. Hence, the gases are thought to contribute to global warming (see this book, section 3.3.2.3).

The reason that these gases have such an effect, whereas other, much more abundant gases such as nitrogen and oxygen do not, is due to their structures and the bonds they contain. The vibrational energies of the O—H bonds in water, the C=O bonds in carbon dioxide and the C—H bonds in methane correspond to energies in the infrared region. When infrared radiation passes through these gases, energy is absorbed (see Table 15) and these bonds stretch and the bond angles in the molecules change (bending). When collisions between molecules occur, the energy absorbed by these molecules can be a transferred to other molecules as, for example, kinetic energy. This process results in an increase in the average energy of the molecules in the atmosphere and, hence, global warming.

Carbon dioxide is the most effective absorber of infrared radiation but, as it comprises only a small percentage of the molecules in the atmosphere, it is not the most important. Water vapour, present in far larger amounts, is responsible for most of the infrared energy absorbed.

Practical and mathematical skills

In the AS paper 2, approximately 15% of marks will be allocated to the assessment of skills related to practical chemistry. A minimum of 20% of the marks will be allocated to assessing level 2 mathematical skills. These practical and mathematical skills are likely to overlap.

The required practical activities assessed in this paper are:

- Investigation of how the rate of a reaction changes with temperature

- Distillation of a product from a reaction

- Tests for alcohol, aldehyde, alkene and carboxylic acid.

The practical skills assessed in the paper are:

1. Independent thinking

Examination questions may require problem solving and the application of scientific knowledge and understanding in practical contexts. For example, a question may ask how, in a novel context, an experiment could be carried out to determine how the rate of an unfamiliar reaction changes with temperature. Another example is a question that requires the evaluation of results from a kinetics experiment.

2. Use and application of scientific methods and practices

This skill may be assessed by asking for critical comments on a given experimental method. Questions may ask for conclusions from given observations: for example, in the reactions of alcohols, aldehydes, alkenes and carboxylic acids with the reagents bromine, Fehling's solution, Tollens' reagent, acidified potassium dichromate(VI) and sodium carbonate. Different methods for the measurement of rates of reaction include an initial rate method, such as a clock reaction and also a continuous monitoring method. Questions on how the rate of a reaction changes with temperature may require the presentation of data in appropriate ways, such as in tables or graphs. It will also be necessary to identify variables, including those that must be controlled, and express numerical results to an appropriate precision with reference to uncertainties and errors in the measurement of time and in thermometer readings.

3. Numeracy and the application of mathematical concepts in a practical context

There is some overlap between this skill and the use and the application of scientific methods and practices. Questions may require the plotting and interpretation of graphs. For example, in experiments to determine the qualitative effect of changes in temperature on the rate of a reaction and the use of Maxwell–Boltzmann distributions to explain why a small temperature increase can lead to a large increase in the rate of a reaction.

4. Instruments and equipment

It will be necessary to know and understand how to set up glassware for distillation and heating under reflux. Methods of heating, including the use of water baths, electrical heaters or sand baths, should be known and their associated hazards understood. Questions will assess the ability to understand in detail how to make appropriate observations of test-tube reactions and how to draw valid conclusions from these observations.

The mathematical skills assessed in this paper are:

1. **Arithmetic and numerical computation**

   - **Recognise and make use of appropriate units in calculations.**

     All numerical answers should be given with the appropriate units. Questions may require conversions between units: for example, cm^3 to dm^3 and J to kJ.

   - **Recognise and use expressions in decimal and standard form.**

     When required, it will be necessary to express answers to an appropriate number of decimal places and to carry out calculations and express answers in ordinary or standard form. For example, calculations involving concentrations may involve numbers in standard form and conversions between standard and ordinary form.

   - **Use ratios fractions and percentages.**

     Examples of this skill include the calculation of percentage yields, atom economies and the construction and/or balancing of equations using ratios.

   - **Estimate results.**

     Calculations of this type could include the evaluation of how a change in concentration of one component in an equilibrium mixture might affect the yield of product.

   - **Use calculators.**

     The ability to use calculators to handle numbers in standard form may be assessed.

2. **Handling data**

   - **Use an appropriate number of significant figures.**

     Understand that a calculated result can only be reported to the limits of the least accurate measurement: for example, the measurement of temperature in an enthalpy of combustion experiment.

   - **Find arithmetic means.**

     Examples may include the determination of the mean bond enthalpy from data for a given bond enthalpy in a range of compounds.

   - **Identify uncertainties in measurements and when data are combined.**

     It will be necessary to demonstrate an ability to determine uncertainty when two readings are used to calculate a value: for example, when a temperature difference is calculated from two thermometer readings.

3. **Algebra**

   - **Change the subject of an equation.**

     For example, when the concentration of a product is calculated from an equilibrium constant expression.

4. **Graphs**

   - **Plot two variables from experimental data.**

     Examples of this skill include the plotting of concentration–time and temperature–time graphs from collected or supplied data for rate of reaction experiments and the drawing of an appropriate best-fit curve.

5. Geometry and trigonometry

- **Visualise and represent 2D and 3D forms.**

  Questions may assess the ability to draw and interpret diagrams of chain, position, functional group and stereoisomers of a given molecule.

- **Use angles and shapes in regular 2D and 3D structures.**

  Questions may assess an ability to predict, identify and sketch the shapes of and bond angles in simple molecules with and without lone pairs: for example, C_3H_8, $CH_3CH{=}CHCH_3$, CH_3OH, $C_2H_5NH_2$, CH_3CHO.

Practice exam-style questions

1 (a) (i) Define the term *enthalpy of combustion*.

_____

_____

_____ 3 marks

(ii) State the condition required for the enthalpy change to be a standard enthalpy change.

_____ 1 mark

(iii) Write an equation for the complete combustion of ethanethiol, CH_3CH_2SH.

_____ 1 mark

(iv) Suggest how the toxic gas formed in this reaction could be removed from the mixture of gases.

_____ 1 mark

(v) Give a reason why the method you have suggested in part (iv) could not be used to remove this gas from the exhaust fumes of a petrol-engined car.

_____ 1 mark

(b) Some standard enthalpies of combustion determined at 298 K are given below. Use these values to find the enthalpy of formation at 298 K of liquid ethanethiol, CH_3CH_2SH.

$$2C(s) + 3H_2(g) + S(s) \rightarrow CH_3CH_2SH(l)$$

	C(s)	$H_2(g)$	S(s)	$CH_3CH_2SH(l)$
$\Delta H^{\ominus}$/kJ mol^{-1}	−394	−286	−297	−2173

_____

_____

_____ 3 marks

_____

Total marks: 10

2 (a) Define the term *activation energy*.

_____

_____ 2 marks

(b) The figure below is a Maxwell–Boltzmann curve for a sample of one mole of oxygen at a given temperature, T_1.

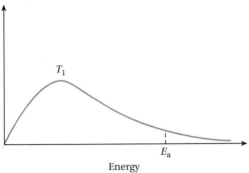

Energy

(i) Label the *y*-axis. 1 mark

(ii) Add a second curve to show the same sample of gas but at a *lower temperature*. Label this curve T_2. 2 marks

(iii) Explain why a small decrease in temperature leads to a much slower rate of reaction.

_____

_____

_____

_____

_____

_____

_____ 3 marks

(c) **(i)** Add and label point X on the correct axis to show the activation energy of the same reaction in the presence of a catalyst. 1 mark

(ii) Explain why the addition of a catalyst increases the rate of a reaction.

_____

_____

_____ 2 marks

(d) State how the Maxwell–Boltzmann curve would differ, if at all, for one mole of hydrogen at the same temperature.

_____ 1 mark

Total marks: 12

3 In a student's experiment, dilute hydrochloric acid and an excess of small marble chips were reacted together and the volume of carbon dioxide evolved was measured at regular time intervals.

The results of one experiment using 10.0 cm³ of hydrochloric acid of concentration 1.00 mol dm⁻³ at 293 K are given below.

Time/s	0	30	60	90	120	150	180	210	240	270	300	330	360	390	420
Volume of CO_2/cm³	0	38	62	77	87	89	97	100	100	102	103	103	104	104	104

(a) Label the axes, plot a graph of the results, and draw a best-fit curve. 4 marks

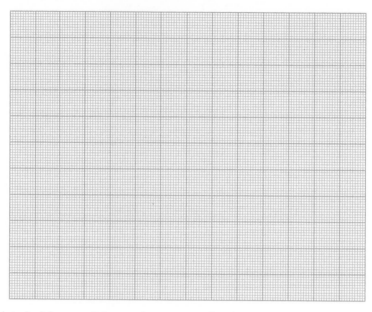

(b) Use your graph to find the rate of the reaction at 90 s and include units in your answer. Show how you have reached your answer on the graph and in your working.

_____

_____

_____

_____ 3 marks

(c) Sketch a curve on your graph to show the results for an experiment carried out with 5.0 cm³ of hydrochloric acid of concentration 2.00 mol dm⁻³ at 293 K, other conditions being unchanged. 2 marks

(d) (i) A student suggested that the percentage uncertainty in measuring the total volume of gas produced from the hydrochloric acid was greatest in measuring the volume of the solution of hydrochloric acid. Use the data to decide whether the student was correct. Suggest how this percentage uncertainty could be reduced and justify your answer.

The solution of hydrochloric acid was measured with a measuring cylinder with an uncertainty of ± 0.5 cm³.

The volume of gas was measured with a gas syringe with a total uncertainty of ±1.0 cm³.

_____

_____

_____

_____

_____

_____

_____ 4 marks

(ii) Explain briefly why the measurement of the mass of the marble chips does not affect the percentage uncertainty in the determination.

_____

_____ 2 marks

Total marks: 15

4 (a) Hydrogen can be made industrially by the reaction of steam with methane. The equation for the reaction is:

$$CH_4(g) + H_2O(g) \rightleftharpoons CO(g) + 3H_2(g) \qquad \Delta H^\ominus = +207 \text{ kJ mol}^{-1}$$

Using le Chatelier's principle, deduce the conditions which would give the maximum yield of hydrogen in this process. Explain how you reach your answer and give reasons why these conditions would not be used in practice.

_____

_____

_____

_____

_____

_____

_____

_____

_____

_____

_____

_____

_____ 6 marks

(b) Consider the equilibrium between carbon monoxide and nitrogen monoxide shown below:

$$2CO(g) + 2NO(g) \rightleftharpoons 2CO_2(g) + N_2(g) \qquad \Delta H^{\ominus} = -746 \text{ kJ mol}^{-1}$$

In one experiment, 0.68 mol of CO is added to 0.58 mol of NO. The mixture is left to reach equilibrium at a given temperature. At equilibrium, the mixture contains 0.33 mol of NO.

The total volume of the equilibrium mixture is 2.00 dm^3.

(i) Deduce the amount, in moles, of CO, CO$_2$ and N$_2$ in the equilibrium mixture.

Moles of CO _____ 1 mark

Moles of CO$_2$ _____ 1 mark

Moles of N$_2$ _____ 1 mark

(ii) Calculate the value of K_c for this equilibrium and give your answer to the appropriate number of significant figures. State the units of K_c.

_____

_____

_____

_____

_____ 4 marks

(c) Deduce the effect on the value of K_c for the equilibrium in part (b) of:

(i) adding a platinum catalyst

_____ 1 mark

(ii) allowing the mixture to reach equilibrium at a higher temperature

_____ 1 mark

Total marks: 15

5 Octane and decane are found in petrol.

(a) An isomer, **Y**, of octane is shown below. Give the IUPAC name of **Y** and state its empirical formula.

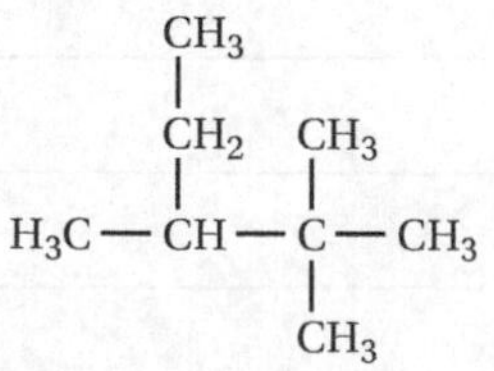

$$
\begin{array}{c}
CH_3 \\
| \\
CH_2 \quad CH_3 \\
| \qquad | \\
H_3C - CH - C - CH_3 \\
| \\
CH_3
\end{array}
$$

IUPAC name _____ 1 mark

Empirical formula _____ 1 mark

(b) Write an equation for the incomplete combustion of decane, the alkane with ten carbon atoms in its molecules, which gives a toxic gas and one other compound.

_____ 1 mark

(c) Carbon is formed in the incomplete combustion that occurs in diesel engines. State why this carbon is a pollutant.

_____ 1 mark

(d) Nitrogen oxides are formed in internal combustion engines.

(i) Explain how nitrogen oxides are formed.

_____

_____ 2 marks

(ii) Identify a metal which is found in a catalytic converter.

_____ 1 mark

(iii) Write an equation to show how nitrogen monoxide can be removed from exhaust gases by reaction with decane to give carbon dioxide, water and nitrogen.

_____ 1 mark

(e) Longer-chain hydrocarbons can be cracked.

(i) Explain what is meant by the term *thermal cracking* and explain in economic terms why thermal cracking is carried out in industry.

_____

_____

_____ 2 marks

(ii) Give an equation for the cracking of dodecane, $C_{12}H_{26}$, to give one molecule of hexane and two molecules of an alkene.

_____ 1 mark

(iii) Identify a reagent which could be used to distinguish between the products of the reaction in part (ii) and state the observations you would make.

_____

_____

_____ 3 marks

Total marks: 14

6 Mechanistically, fluorine reacts with ethane in the same way as chlorine.

(a) Fluorine reacts with ethane to form fluoroethane.

(i) Name the mechanism.

_____ 1 mark

(ii) Write an equation for the initiation step in this reaction. State the essential condition for this step.

Equation _____ 1 mark

Essential condition _____ 1 mark

(iii) Write equations for the two propagation steps.

Equation 1 _____ 1 mark

Equation 2 _____ 1 mark

(iv) Write an equation for the termination step in which a molecule that has the empirical formula C_2H_5 is formed.

_____ 1 mark

(b) 1,1,1-Trifluoroethane is one of a series of compounds known as HFCs which can be used as replacements for CFCs in refrigerants and aerosols.

(i) Write an equation for the reaction of fluorine with ethane to make 1,1,1-trifluoroethane showing the structure of the product as a displayed formula.

1 mark

(ii) Explain how 1,1,1-trichloroethane contributes to the depletion of the ozone layer. In your answer, identify the catalyst and explain how it is acting as the catalyst.

Give equations in your answer for any relevant reactions.

_____

_____

_____

_____

_____

_____ 7 marks

(iii) Explain why 1,1,1-trifluoroethane does not contribute to the destruction of the ozone layer.

_____

_____ 2 marks

Total marks: 16

7   Chemists often wish to synthesise new molecules with longer carbon chains than the starting molecule. One method of adding another carbon atom to a chain uses cyanide ions.

(a) Write an equation for the reaction of bromoethane with potassium cyanide and name the organic product.

_____

_____ 2 marks

(b) Name and outline a mechanism for this reaction.

Name of mechanism _____

Mechanism

3 marks

(c) Predict, with a reason, whether the reaction of chloroethane with potassium cyanide would be faster or slower than the reaction in part (a).

_____

_____

_____ 2 marks

Total Marks: 7

8   **(a)** Analysis of an organic liquid, **X**, shows that it contains, by mass, 54.54% carbon; 9.09% hydrogen; and the rest is oxygen.

Calculate the empirical formula of the organic liquid.

_____

_____

_____

_____

_____

_____

_____

_____

_____ 3 marks

(b) The mass spectrum of **X** is shown below. Use it to find the molecular mass of **X** and hence deduce the molecular formula of **X**.

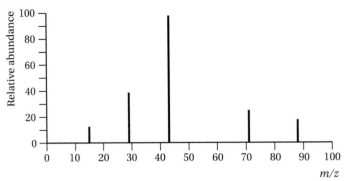

Molecular mass _____ 1 mark

Molecular formula _____ 1 mark

Total Marks: 5

9 (a) (i) Give the full IUPAC name of the molecule shown below:

IUPAC name _____ 1 mark

(ii) Draw the skeletal formula of a functional group isomer of pent-1-ene.

1 mark

(iii) Draw the displayed formula of a positional isomer of pent-1-ene and give the full IUPAC name of your structure.

1 mark

(iv) Draw the displayed formula of a chain isomer of pent-1-ene and give the IUPAC name of your structure.

1 mark

(b) Alkenes can be formed from both halogenoalkanes and from alcohols.

(i) Give a reagent and the necessary conditions for the conversion of 1-bromobutane into but-1-ene. Write an equation for the reaction.

Reagent _____ 1 mark

Conditions _____ 1 mark

Equation _____ 1 mark

(ii) Name and outline the mechanism for the dehydration reaction of butan-2-ol in the presence of concentrated sulfuric acid to give but-1-ene.

5 marks

(c) Hydrogen bromide, HBr, reacts with but-1-ene to produce a pair of structural isomers.

Name and outline the mechanism of the reaction to form the major isomer and explain why this is the major product.

Explanation _____

_____

_____ 7 marks

Total Marks: 19

10 (a) Draw the repeating unit of the polymer formed by ethene and the repeating unit of the polymer formed by propene. Name the type of polymerisation involved.

Repeating unit from ethene _____

Repeating unit from propene _____

Type of polymerisation _____ 3 marks

(b) Explain why these two polymers have relatively low melting points.

_____

_____ 2 marks

(c) Explain why it makes commercial sense to recycle objects made from poly(propene) rather than to bury them as household waste.

_____

_____ 2 marks

Total Marks: 7

11 (a) Write an equation to show the formation of ethanol by fermentation of glucose and give the conditions used for this reaction.

Equation _____

Conditions _____

_____ 4 marks

(b) Ethanol is an important solvent in industrial processes. Discuss the advantages and disadvantages of the production of ethanol by fermentation compared with its production by the direct hydration of ethene.

_____

_____

_____

_____ 4 marks

(c) Explain what is meant by the term *biofuel*.

_____ 1 mark

Total Marks: 9

12 (a) Write an equation for the complete combustion of ethanol.

_____ 1 mark

(b) Explain why the use of ethanol as a fuel can be considered to be carbon neutral.

_____

_____

_____

_____

_____

_____ 6 marks

Total Marks: 7

13 The structures of two isomeric alcohols are shown below:

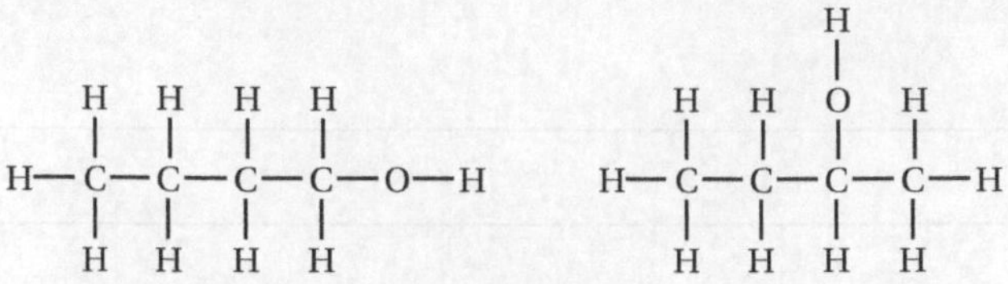

$$
\begin{array}{cccc}
& H & H & H & H \\
& | & | & | & | \\
H-C & -C & -C & -C & -O-H \\
& | & | & | & | \\
& H & H & H & H \\
\end{array}
\qquad
\begin{array}{cccc}
& & H & & \\
& & | & & \\
& H & O & H & \\
& | & | & | & \\
H-C & -C & -C & -C-H \\
& | & | & | & | \\
& H & H & H & H \\
\end{array}
$$

A **B**

(a) Identify the type of structural isomerism shown by **A** and **B** and give the class of alcohols to which each belongs.

Type of isomerism _____ 1 mark

Class of alcohol for **A** _____ 1 mark

Class of alcohol for **B** _____ 1 mark

(b) There are two further isomeric alcohols of **A** and **B** with branched chains. Draw their displayed structures and give their IUPAC names. Classify each isomer.
Isomer **C**

_____ 1 mark

IUPAC name _____ 1 mark

Class of alcohol _____ 1 mark

Isomer **D**

_____ 1 mark

IUPAC name _____ 1 mark

Class of alcohol _____ 1 mark

(c) One of the isomers **A** to **D** cannot be oxidised by acidified potassium dichromate(VI) solution.

Identify this isomer and suggest why this isomer cannot be oxidised.

Isomer _____ 1 mark

Reason _____ 1 mark

(d) One of the isomers can be oxidised to a ketone.

Identify this isomer and write an equation for the reaction. Use [O] to represent the oxidising agent.

Isomer _____ 1 mark

Equation _____ 1 mark

(e) The remaining two alcohols can both be oxidised to aldehydes and carboxylic acids by acidified potassium dichromate(VI) solution.

Explain how you could control the conditions of the preparation to maximise the yield of the aldehyde and to maximise the yield of the carboxylic acid in different experiments. In your answer, explain why these different conditions maximise the yield.

_____

_____

_____

_____

_____

_____

_____

_____

_____

_____

_____

_____

_____

_____ 6 marks

Total Marks: 19

14 (a) Analysis of an organic compound, **X**, showed that its relative molecular mass was 102. This value could arise from several different molecular formulae, including:

(i) $C_4H_6O_3$, (ii) $C_5H_{10}O_2$ and (iii) $C_3H_6N_2O_2$

Analysis of the high-resolution mass spectrum of **X** showed that the molecular ion of this compound produces a peak at $m/z = 102.1313$. Use the precise relative atomic masses given below to deduce which one of the three molecular formulae corresponds to **X** and show that the other two do not.

C = 12.0107   H = 1.0079   O = 15.9994   N = 14.0067

_____

_____

_____ 3 marks

(b) The infra-red spectrum of **X** is shown below.

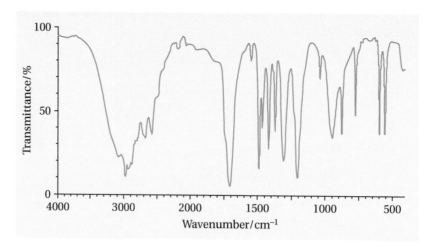

(i) Use the table of infra-red absorptions provided below to identify the functional group present in **X**.

_____ 1 mark

(ii) Draw the structures of the isomers of **X** which contain this functional group.

4 marks

(iii) Describe how the fingerprint region of the infra-red spectrum could be used to identify which isomer was **X**.

_____

_____ 2 marks

Total Marks: 10

Multiple-choice questions

1   An organic liquid is vaporised at 120 °C. 0.34 g of the liquid occupied a volume of 191 cm^3 at a pressure of 100 kPa. The M_r of the liquid is:

   A   18

   B   58

   C   171

   D   5816

2   A skeletal structure of a molecule, **X**, is shown:

   Choose the letter which corresponds to the incorrect statement about **X**:

   A   **X** undergoes electrophilic addition.

   B   The full IUPAC name for **X** is Z-1-bromo-2-methylbut-2-ene.

   C   **X** undergoes free-radical substitution.

   D   **X** undergoes nucleophilic substitution.

3   Choose the letter which corresponds to the incorrect statement:

   A   Propan-1-ol has a higher boiling temperature than butane.

   B   Fehling's reagent can distinguish butanal from butanone.

   C   Butanal has a strong absorption in the infrared between 1680 and 1750 cm^{-1}

   D   2-methylbutan-2-ol can be oxidised by acidified potassium dichromate(VI)

4   Choose the letter which corresponds to the incorrect statement:

   A   The area under a Maxwell–Boltzmann distribution curve for a fixed amount of gas does not depend on the temperature.

   B   The XeF$_4$ molecule has bond angles of 90°.

   C   Heating a reaction in the gas phase increases the activation energy of the particles.

   D   A catalyst increases the rate of attainment of an equilibrium but it does not affect the value of the equilibrium constant, K_c for the reaction.

Answers

Question	Answer		Marks
1 (a) (i)	enthalpy change when one mole of a substance	(1)	
	is completely burnt in oxygen	(1)	
	with all reactants and products in their standard states	(1)	3
1 (a) (ii)	100 kPa pressure	(1)	1
1 (a) (iii)	$CH_3CH_2SH + 4\frac{1}{2}O_2 \rightarrow 2CO_2 + 3H_2O + SO_2$	(1)	1
1 (a) (iv)	react with calcium oxide	(1)	1
1 (a) (v)	reactions too slow/calcium oxide would need replacing too often	(1)	1
1 (b)	$\Delta_f H = \Sigma\Delta_c H(\text{reactants}) - \Sigma\Delta_c H(\text{products})$	(1)	
	$= [(2 \times -394) + (3 \times -286) + -297)] - (-2173)$	(1)	
	$= +230 \text{ kJ mol}^{-1}$	(1)	3
			Total 10
2 (a)	energy needed for a successful collision	(1)	
	minimum energy needed	(1)	2
2 (b) (i)	number of molecules	(1)	1
2 (b) (ii)			
	curve has higher peak and peak to the left of first curve	(1)	
	curve starts at origin, tends to x-axis below first curve	(1)	2
2 (b) (iii)	fewer molecules have $E \geq E_a$	(1)	
	a small decrease in temperature means **many** fewer molecules have $E \geq E_a$	(1)	
	so many fewer collisions with $E \geq E_a$	(1)	3
2 (c) (i)			
		(1)	1

Question	Answer		Marks
2 (c) (ii)	provides an alternative pathway with lower E_a	(1)	
	so many more molecules have $E \geq E_a$ and more collisions are successful	(1)	2
2 (d)	no difference	(1)	1
			Total 12
3 (a)	x-axis labelled time/s and y-axis labelled volume (of CO_2)/cm^3 with values that use more than half of each axis	(1)	
	all points plotted correctly	(1)	
	smooth curve drawn	(1)	
	ignoring anomalous points at 150s / 89 cm^3 and 240 s / 100 cm^3	(1)	4
3 (b)	gradient drawn at 90 s and triangle drawn	(1)	
	value 0.40 (0.36 – 0.44)	(1)	
	units $cm^3 \, s^{-1}$	(1)	3
3 (c)	curve starting at the origin with greater initial gradient	(1)	
	ending at same value of 104 cm^3 to within ± ½ small square	(1)	2
3 (d) (i)	percentage uncertainty in measuring solution = 0.5 / 10.0 × 100 = 5.0%	(1)	
	percentage uncertainty in measuring volume = 1.0 / 104.0 × 100 = 0.96%	(1)	
	so the student was correct	(1)	
	use a pipette/burette to measure the solution	(1)	4
3 (d) (ii)	the marble chips are in excess	(1)	
	the exact mass of chips does not affect the final volume providing the chips are in excess	(1)	2
			Total 15
4 (a)	**This answer is marked using levels of response.**		
	Level 3: 5–6 marks		
	All parts are covered and the explanation of each part is generally correct and virtually complete.		
	Answer communicates the whole process coherently and shows a logical progression from part 1 and part 2 to overall reasons.		
	Level 2: 3–4 marks		
	All parts are covered but the explanation of each part may be incomplete OR two parts are covered and the explanations are virtually complete.		
	Answer is mainly coherent and shows a progression. Some statements may be out of order and incomplete.		
	Level 1: 1–2 marks		
	Two parts are covered but the explanation of each part may be incomplete and contain inaccuracies OR only one part is covered but the explanation is mainly correct and is virtually complete.		
	Answer includes some isolated statements but there is no attempt to present them in a logical order or show confused reasoning.		
	Level 0: 0 marks		
	Insufficient correct chemistry to warrant a mark.		

Question	Answer		Marks
	Effect of temperature: reaction is endothermic so higher temperatures move equilibrium to the right ∴ high temperature would give maximum yield		
	Effect of pressure: more moles of gas on right so lower pressures move equilibrium to the right ∴ low pressure would give maximum yield		
	Reasons why: higher temperatures require high energy input to maintain and therefore have high fuel costs at low pressures particles have fewer collisions so rate would be slow compromise between yield and rate and cost would give moderately high temperature and moderately low pressure as optimum conditions		
4 (b) (i)	moles of CO: 0.43 moles of CO_2: 0.25 moles of N_2: 0.125	(1) (1) (1)	3
4 (b) (ii)	$K_c = [CO_2]^2[N_2]/[CO]^2[NO]^2$ $= (0.25 / 2)^2 (0.125 / 2) / (0.43 / 2)^2 (0.33 / 2)^2$ $= 0.775\,988 = 0.78$ (2 sig. figs.) units: $dm^3\,mol^{-1}$	(1) (1) (1) (1)	4
4 (c) (i)	no change	(1)	1
4 (c) (ii)	K_c decreases	(1)	1
			Total 15
5 (a)	*IUPAC name* 2,2,3-trimethylpentane *Empirical formula* C_4H_9	(1) (1)	2
5 (b)	$C_{10}H_{22} + 10\frac{1}{2}O_2 \rightarrow 10CO + 11H_2O$	(1)	1
5 (c)	forms smog/causes bronchitis	(1)	1
5 (d) (i)	at the high temperatures in internal combustion engines nitrogen and oxygen react together	(1) (1)	2
5 (d) (ii)	palladium/platinum/iridium/rhodium OR Pt/Pd/Ir/Rh	(1)	1
5 (d) (iii)	$31NO + C_{10}H_{22} \rightarrow 10CO_2 + 11H_2O + 15\frac{1}{2}N_2$	(1)	1
5 (e) (i)	longer chain alkanes are broken into shorter alkanes and alkenes by breaking carbon–carbon bonds in the chain shorter chain alkanes are more useful as fuels/alkenes are more chemically reactive so have more value	(1) (1)	2
5 (e) (ii)	$C_{12}H_{26} \rightarrow C_6H_{14} + 2C_3H_6$	(1)	1
5 (e) (iii)	bromine water alkanes: no visible change alkenes: decolourises the bromine water/turns from orange to colourless	(1) (1) (1)	3
			Total 14
6 (a) (i)	free-radical substitution	(1)	1

Question	Answer		Marks
6 (a) (ii)	*Equation* $F_2 \rightarrow 2F\bullet$	(1)	
	Essential condition uv light	(1)	2
6 (a) (iii)	*Equation 1* $F\bullet + CH_3CH_3 \rightarrow HF + \bullet CH_2CH_3$	(1)	
	Equation 2 $F_2 + \bullet CH_2CH_3 \rightarrow F\bullet + FH_2CCH_3$	(1)	2
6 (a) (iv)	$2 \bullet CH_2CH_3 \rightarrow CH_3CH_2CH_2CH_3$	(1)	1
6 (b) (i)			

$$3F_2 \; + \; H-\underset{\underset{H}{|}}{\overset{\overset{H}{|}}{C}}-\underset{\underset{H}{|}}{\overset{\overset{H}{|}}{C}}-H \longrightarrow F-\underset{\underset{F}{|}}{\overset{\overset{F}{|}}{C}}-\underset{\underset{H}{|}}{\overset{\overset{H}{|}}{C}}-H \; + \; 3HF$$

		(1)	1
6 (b) (ii)	C—Cl bond breaks when uv light absorbed	(1)	

$$Cl-\underset{\underset{Cl}{|}}{\overset{\overset{Cl}{|}}{C}}-\underset{\underset{H}{|}}{\overset{\overset{H}{|}}{C}}-H \longrightarrow \bullet\underset{\underset{Cl}{|}}{\overset{\overset{Cl}{|}}{C}}-\underset{\underset{H}{|}}{\overset{\overset{H}{|}}{C}}-H \; + \; Cl\bullet$$

		(1)	
	to form $Cl\bullet$ which is the catalyst	(1)	
	Equation 1 $Cl\bullet + O_3 \rightarrow Cl-O\bullet + O_2$	(1)	
	Equation 2 $Cl-O\bullet + O_3 \rightarrow Cl\bullet + 2O_2$	(1)	
	Overall $2O_3 \rightarrow 3O_2$	(1)	
	$Cl\bullet$ is the catalyst because it provides an alternative route	(1)	
	and is regenerated at the end	(1)	
	(any 7 of the above)		7
6 (b) (iii)	has no C—Cl bonds to break so cannot form $Cl\bullet$	(1)	
	C—F bonds too strong to break so $F\bullet$ not formed	(1)	2
			Total 16
7 (a)	$CH_3CH_2Br + CN^- \rightarrow CH_3CH_2CN + Br^-$	(1)	
	propanenitrile	(1)	2
7 (b)	nucleophilic substitution	(1)	
		(1)	
	$NC\overset{-}{:}$ $CH_3CH_2 \underset{\delta+}{} Br^{\delta-} \longrightarrow CH_3CH_2CN \; + \; :Br^-$	(1)	3
7 (c)	reaction with chloroethane will be slower	(1)	
	C—Cl bonds are stronger than C—Br	(1)	2
			Total 7

8 (a)	Element	% by mass	$\div A_r$		/2.27			
	C	54.54	$\div 12.0$	$\rightarrow 4.54$	2			
	H	9.09	$\div 1.0$	$\rightarrow 9.09$	4			
	O	36.37 (1)	$\div 16.0$	$\rightarrow 2.27$ (1)	1			

	so empirical formula is C_2H_4O	(1)	3
8 (b)	*Molecular mass* 88.0	(1)	
	Molecular formula $C_4H_8O_2$	(1)	2
			Total 5

Question	Answer	Marks
9 (a) (i)	*full IUPAC name* Z-3-methylhex-2-ene (1)	1
9 (a) (ii)	*etc.* (1)	1
9 (a) (iii)	 Z-pent-2-ene or E-pent-2-ene (1)	1
9 (a) (iv)	 2-methylbut-1-ene or 3-methylbut-1-ene (1)	1
9 (b) (i)	*Reagent* NaOH (1) *Conditions* ethanol/heat (1) *Equation* $CH_3CH_2CH_2CH_2Br + NaOH \rightarrow CH_3CH_2CH{=}CH_2 + NaBr + H_2O$ (1)	3
9 (b) (ii)	Name of mechanism: acid-catalysed elimination (1) 1 mark for each curly arrow (3); 1 mark for intermediate structure with +ve charge on O (1)	5
9 (c)	Name of mechanism: electrophilic addition (1) One mark for each curly arrow (3); one for carbocation intermediate (1) *Explanation* the two possible intermediates are primary/1° carbocations and secondary/2° carbocations (1) secondary carbocations are more stable than primary so react to give the major product (1)	7
		Total 19

Question	Answer	Marks
10 (a)	$$\begin{array}{ccc} & H & H \\ & | & | \\ -C & - & C- \\ & | & | \\ & H & H \end{array}$$ (1) $$\begin{array}{ccc} & CH_3 & H \\ & | & | \\ -C & - & C- \\ & | & | \\ & H & H \end{array}$$ (1) addition polymerisation (1)	3
10 (b)	weak van der Waals' forces between polymer chains (1) little energy required to overcome them (1)	2
10 (c)	polyalkenes are not biodegradable (1) poly(propene) can be remelted and remoulded into new objects (1)	2
		Total 7
11 (a)	$C_6H_{12}O_6 \rightarrow 2C_2H_5OH + 2CO_2$ (1) yeast (1) aqueous solution (1) 35 °C (1)	4
11 (b)	(see table below)	4
11 (c)	fuel made from biomass (1)	1
		Total 9
12 (a)	$C_2H_5OH + 3CO_2 \rightarrow 2CO_2 + 3H_2O$ (1)	1
12 (b)	fermentation of one mole of glucose forms two moles of CO_2 together with two moles of ethanol (1) combustion of these two moles of ethanol produces another four moles of CO_2 (1) photosynthesis uses six moles of CO_2 to produce one mole of glucose (1) so there is no net production of CO_2 (1) but there will be energy considerations in the production of ethanol and its use (1) so there will be some overall carbon dioxide production (1)	6
		Total 7
13 (a)	*Type of isomerism* positional (1) *Class of alcohol for A* primary/1° (1) *Class of alcohol for B* secondary/2° (1)	3

11 (b) table:

Method	Rate of reaction	Quality of product	Raw material	Type of process
hydration	fast	pure	ethene from oil (finite resource)	continuous (cheap on manpower) (expensive equipment)
fermentation	slow (1)	impure (1)	sugars (renewable resource) (1)	batch (expensive on manpower) (cheap equipment) (1)

Question	Answer		Marks
13 (b)	Either order Isomer **C** 	(1)	
	IUPAC name 2-methylpropan-2-ol	(1)	
	Class of alcohol 3°/tertiary	(1)	3
	Isomer **D** 	(1)	
	IUPAC name 2-methylpropan-1-ol	(1)	
	Class of alcohol 1°/primary	(1)	3
13 (c)	Isomer 2-methylpropan-2-ol (isomer **C** above)	(1)	
	Reason does not have a hydrogen atom on the carbon with the OH group	(1)	2
13 (d)	Isomer **B**/butan-2-ol	(1)	
	Equation 	(1)	2
13 (e)	Aldehyde: add oxidising agent slowly to alcohol	(1)	
	distil off the aldehyde as soon as it is formed	(1)	
	to prevent further oxidation	(1)	3
	Carboxylic acid: use an excess of oxidising agent	(1)	
	heat the mixture under reflux	(1)	
	to ensure complete oxidation	(1)	3
			Total 19
14 (a) (i)	$C_4H_6O_3 = 102.0315$	(1)	1
14 (a) (ii)	$C_5H_{10}O_2 = 102.0678 = \mathbf{X}$	(1)	1
14 (a) (iii)	$C_3H_6N_2O_2 = 102.0428$	(1)	1

Question	Answer	Marks
14 (b) (i)	carboxylic acid (1)	1
14 (b) (ii)	$CH_3CH_2CH_2CH_2COOH$ (1)　　　　CH_3CHCH_2COOH (1) with CH_3 branch $CH_3CH_2CHCOOH$ (1) with CH_3 branch　　$H_3C—C—COOH$ (1) with two CH_3 branches	4
14 (b) (iii)	compare the fingerprint region of the infra-red spectrum of **X** with those of the four acids (1) an exact match will indicate which one is **X** (1)	2
		Total 10

Multiple-choice questions

1 B	
2 B	
3 D	
4 C	

The table below highlights aspects of *mathematical and practical skills* in the exemplar questions.

Question	Mathematical skill
1 b	MS 2.4
2 b	MS 3.1
2 c	MS 3.1
3 a	MS 3.2
3 b	MS 0.0, MS 3.5
3 c	MS 3.1
3 d	MS 0.2, MS 1.3
4 b i	MS 0.2
4 b ii	MS 0.0, MS 0.1, MS 1.1, MS 2.1, MS 2.3, MS 2.4
4 c	MS 3.1
5 b	MS 0.2
5 d iii	MS 0.2
5 e ii	MS 0.2
6 a	MS 0.2
8 a	MS 0.2
8 b	MS 3.1
9 a i	MS 4.1, MS 4.2

9 a iii and iv	MS 4.2
10 a	MS 4.2
13	MS 4.2
14	MS 3.1
MCQ 1	MS 0.0, MS 2.2
MCQ 2	MS 4.2, MS 4.3
MCQ 4	MS 4.1, MS 4.3

Question	Practical skill
3 a	PS 1.1
3 b	PS 3.1
3 c	PS 3.2
3 d	PS 2.3, PS 3.3
9 c	PS 1.2
13 e	PS 4.1
14	PS 1.1

Glossary

acid rain	contains quantities of carbonic, nitric and sulfuric acids
activation energy	the minimum energy required for a reaction to occur
addition polymer	one obtained by the addition of monomers to the end of a growing chain
allotropes	different structural modifications of an element
atomic (proton) number (Z)	the number of protons in the nucleus of an atom
Avogadro constant (L)	6.022×10^{23} mol^{-1}
backward (or reverse) reaction	one that goes from right to left in an equation
biofuel	fuel produced from renewable plant material (biomass)
bond dissociation enthalpy	the enthalpy change for the breaking of a covalent bond, with all species in the gaseous state
Cahn–Ingold–Prelog (CIP) priority rules	used to distinguish between stereoisomers
calorimeter	apparatus used to measure heat change
carbocation	a species which contains a carbon atom that has a positive charge
carbon neutral	applies to a process which occurs without any change in the total amount of carbon dioxide present in the atmosphere
catalyst	a substance which alters the rate of a reaction without itself being consumed
catalytic cracking	occurs when the energy required for bond breaking in hydrocarbons is provided by heat, in the presence of a catalyst (compare with *thermal cracking*)
chain isomers	structural isomers which occur when there are two or more ways of arranging the carbon skeleton of a molecule
chain reaction	one in which many molecules undergo chemical reaction after one molecule becomes activated
chemical equilibrium	the point at which, in a reversible reaction, both the forward and backward reactions occur at the same rate, with the concentrations of all reactants and products remaining constant
concentration	$\dfrac{\text{number of moles of solute}}{\text{volume of solution in dm}^3}$ with units mol dm^{-3}
co-ordinate bond	a covalent bond formed when the pair of electrons originate from one atom
covalent bond	a shared pair of electrons
cracking	occurs when large alkanes are broken into smaller molecules
dative covalent bond	a covalent bond formed when the pair of electrons originates from one atom
delocalised electrons	electrons that are spread over many ions in a metal and which are free to move through the lattice

dispersion forces	the weakest forces of attraction that exist between atoms or molecules, which results when electrons on adjacent atoms are displaced and induce temporary dipoles (also known as van der Waals' forces or London forces)
displayed formula	shows all the bonds present in a molecule
dynamic reaction	one which proceeds simultaneously in both directions
E–Z stereoisomerism	also known as geometrical or *cis–trans* isomerism
E–Z stereoisomers	isomers that arise due to restricted rotation about a carbon–carbon double bond when the two pairs of attached substituents can be arranged in two different ways
elastic collisions	those in which no energy is lost on collision
electronegativity	the power of an atom to attract the electrons in a covalent bond
electrophilic addition reaction	one in which a C=C double bond becomes saturated; the mechanism involves initial attack by an electron-deficient species (electrophile)
elimination reaction	one in which an unsaturated compound is formed by the removal of a small molecule such as hydrogen bromide
empirical formula	the simplest ratio of atoms of each element in a compound
endothermic	the gain of heat energy by a system; the enthalpy change is positive
endothermic reaction	one in which heat energy is taken in
enthalpy of fusion	the enthalpy required to change one mole of a solid into a liquid, i.e. $X(s) \rightarrow X(l)$
enthalpy of vaporisation	the enthalpy required to change one mole of a liquid into a gas, i.e. $X(l) \rightarrow X(g)$
equilibrium constant (K_c)	the ratio of concentrations of products and reactants raised to the powers of their stoichiometric coefficients; e.g. for the reaction $3A \rightleftharpoons 2B + C$ $\quad K_c = \dfrac{[B]^2[C]}{[A]^3}$
exothermic	the loss of heat energy by a system; the enthalpy change is negative
exothermic reaction	one in which heat energy is given out
Fehling's solution	contains a deep blue copper(II) complex ion which, with aldehydes (but not ketones), is reduced, on warming, to form a red precipitate of copper(I) oxide
fingerprint region	the region below 1500 cm^{-1} in an infrared spectrum
first law of thermodynamics	energy can neither be created nor destroyed, but can only be converted from one form into another
free-radical substitution reaction	one in which the hydrogen atom of a C—H bond is replaced by a halogen atom; the chain-reaction mechanism involves attack on a neutral molecule by a radical (halogen atom)
forward reaction	one that goes from left to right in an equation
functional group	an atom or group of atoms which, when present in different molecules, causes them to have similar chemical properties

functional group isomers	structural isomers which contain different functional groups
giant ionic lattice	see *ionic crystal*
giant metallic lattice	see *metallic crystal*
greenhouse gases	gases in the atmosphere which absorb infrared radiation (e.g. water vapour, carbon dioxide, methane and ozone)
Hess's law	the enthalpy change of a reaction depends only on the initial and final states of the reaction and is independent of the route by which the reaction occurs
homolytic fission	formation of radicals when a covalent bond breaks with an equal splitting of the bonding pair of electrons
homologous series	a family of organic molecules which all contain the same functional group, but have an increasing number of carbon atoms; each member can be represented by a general formula, e.g. $C_nH_{2n+1}X$
hydrogen bonding	an intermolecular force between the lone pair on an electronegative atom (N, O or F) and a hydrogen atom bonded to such an electronegative atom
ideal gas	one that obeys the ideal gas equation, $pV = nRT$
integral mass	relative molecular mass to the nearest whole number
ion	an atom or group of atoms which has lost or gained one or more electrons, giving it a positive or negative charge
ionic bond	the electrostatic force of attraction between oppositely charged ions
ionic crystal	a lattice of positive and negative ions bound together by electrostatic attractions
isomers	molecules with the same chemical formula but in which the atoms are arranged differently (see *structural isomerism* and *stereoisomerism*)
isotopes	atoms of the same element with the same atomic number but different mass numbers
Le Chatelier's principle	a system at equilibrium will respond to oppose any change imposed upon it
London forces	the weakest forces of attraction that exist between atoms or molecules, which results when electrons on adjacent atoms are displaced and induce temporary dipoles (also known as van der Waals' forces or dispersion forces)
macromolecular (giant) crystal	a large, covalently-bonded lattice structure
macromolecule	a large molecule with a regular three- or two-dimensional lattice structure
mean bond enthalpy	the average of several values of the bond dissociation enthalpy for a given type of bond, taken from a range of different compounds
mechanism	the steps by which a reaction occurs
metallic bonding	electrostatic attraction between metal ions and delocalised electrons
metallic crystal	a lattice of metal ions surrounded by delocalised electrons
mole	the amount of substance of a system which contains as many elementary entities as there are atoms in 0.012 kilogram of carbon-12; the elementary entities must be specified and may be atoms, molecules, ions, electrons, other particles, or specified groups of such particles

molecular crystal	a lattice of covalent molecules held together by weak intermolecular forces
molecular fingerprint	the unique entire infrared spectrum of an organic compound
molecular formula	the actual number of atoms of each element in a molecule
molecular ion ($M^{+\bullet}$)	the species formed in a mass spectrometer by the loss of one electron from a molecule
nucleophilic substitution reaction	one in which an electron-rich molecule or anion (with a lone pair of electrons) attacks an electron-deficient carbon atom, resulting in the replacement of an atom or group of atoms originally attached to this carbon
octahedral	the spatial arrangement with one atom at the centre of six other atoms, with four atoms in its plane, one atom above this plane and one atom below this plane
percentage atom economy	$$\frac{\text{mass of desired product}}{\text{total mass of reactants}} \times 100$$ it is a measure of how much of a desired product in a reaction is formed from the reactants
percentage yield	$$\frac{\text{actual mass of product}}{\text{maximum theoretical mass of products}} \times 100$$ it is a practical measure of the efficiency of a reaction
permanent dipole–dipole force	attraction between the slightly positive end of one polar molecule and the slightly negative end of an adjacent polar molecule
polarity	the displacement of electron density (formation of an electric dipole) in a covalent bond, or in a molecule, due to a difference in electronegativity
position isomers	structural isomers which have the same carbon skeleton and the same functional group, but in which the functional group is joined at different places on the carbon skeleton
rate of reaction	the change in concentration of a substance in unit time
reflux	a process in which a reaction mixture is heated in a flask fitted with a condenser to prevent the loss of volatile substances, including the solvent
relative atomic mass (A_r)	$$\frac{\text{average mass of one atom of an element}}{\frac{1}{12} \times \text{the mass of one atom of } {}^{12}\text{C}}$$
relative molecular mass (M_r)	$$\frac{\text{average mass of one molecule}}{\frac{1}{12} \times \text{the mass of one atom of } {}^{12}\text{C}}$$
reversible reaction	one which does not go to completion but can occur in either direction
saturated hydrocarbons	contain carbon–carbon single bonds as well as carbon–hydrogen bonds
skeletal formula	the most abbreviated type of formula; the basic carbon skeleton is shown by two-dimensional zig-zag lines, without the use of any carbon or hydrogen atoms
square planar	the spatial arrangement of a central atom surrounded by four atoms situated at the corners of a square
standard conditions	usually taken as 100 kPa and 298 K
standard enthalpy of combustion ($\Delta_c H^{\ominus}$)	the enthalpy change, under standard conditions, when 1 mol of a substance is burned completely in oxygen, with all reactants and products in their standard states

standard enthalpy of formation ($\Delta_f H^{\ominus}$)	the enthalpy change, under standard conditions, when 1 mol of a compound is formed from its elements, with all reactants and products in their standard states
standard state	the normal, stable state of an element or compound under standard conditions, usually 298 K and 100 kPa
stereoisomerism	occurs when molecules with the same structural formula have the bonds arranged differently in space
stereoisomers	are compounds which have the same structural formula but have bonds that are arranged differently in space
stoichiometric coefficient	the number of moles of a species as shown in a balanced equation
structural formula	shows the unique arrangement of atoms in a molecule in a simplified form without showing all the bonds
structural isomerism	occurs when the component atoms are arranged differently in molecules having the same molecular formula
structural isomers	compounds with the same molecular formula but different structures
substitution reaction	one in which an atom or functional group in a molecule is replaced by another atom or functional group
tetrahedral	the spatial arrangement with one atom at the centre of a tetrahedron of four other atoms
thermal cracking	occurs when the energy required for bond breaking in hydrocarbons is provided by heat alone (compare with *catalytic cracking*)
Tollens' reagent	contains the complex ion $[Ag(NH_3)_2]^+$ which, with aldehydes (but not ketones), is reduced, on warming, to silver
trigonal bipyramidal	the spatial arrangement with one atom at the centre of five other atoms, with three atoms in its plane, one atom above this plane and one atom below this plane
trigonal planar	the spatial arrangement with one atom at the centre of a triangle of three other atoms, with all four atoms in the same plane
trivial name	a simple, often historical, name used where the systematic name is cumbersome or complicated; for example, β-carotene, found in carrots, has the IUPAC name 1,3,3-trimethyl-2-[(1E,3E,5E,7E,9E,11E,13E,15E,17E)-3,7,12,16-tetramethyl-18-(2,6,6-trimethylcyclohexen-1-yl)octadeca-1,3,5,7,9,11,13,15,17-nonaenyl]cyclohexene
unsaturated	is applied to molecules which contain at least one $C{=}C$ double covalent bond or one $C{\equiv}C$ triple covalent bond
van der Waals' force	the weakest forces of attraction that exist between atoms or molecules, which results when electrons on adjacent atoms are displaced and induce temporary dipoles (also known as dispersion forces or London forces)
wavenumber	reciprocal wavelength ($1/\lambda$) with units cm^{-1}, used to indicate band positions in infrared spectra

Index

Notes

Notes